T0254701

Optimization Modeling Using R

Chapman & Hall/CRC Series in Operations Research

Series Editors:
Malgorzata Sterna, Bo Chen, Michel Gendreau, and Edmund Burke

Rational Queueing
Refael Hassin

Introduction to Theory of Optimization in Euclidean Space
Samia Challal

Handbook of The Shapley Value
Encarnación Algaba, Vito Fragnelli and Joaquín Sánchez-Soriano

Advanced Studies in Multi-Criteria Decision Making
Sarah Ben Amor, João Luís de Miranda, Emel Aktas, and Adiel Teixeira de Almeida

Handbook of Military and Defense Operations Research
Natalie Scala, and James P. Howard II

Understanding Analytic Hierarchy Process
Konrad Kulakowski

Introduction to Optimization-Based Decision Making
João Luís de Miranda

Applied Mathematics with Open-source Software
Operational Research Problems with Python and R
Vince Knight, Geraint Palmer

Optimization Modeling Using R
Timothy R. Anderson

For more information about this series please visit: https://www.routledge.com/Chapman--HallCRC-Series-in-Operations-Research/book-series/CRCOPSRES

Optimization Modeling Using R

Timothy R. Anderson

CRC Press is an imprint of the
Taylor & Francis Group, an **informa** business
A CHAPMAN & HALL BOOK

First edition published 2023
by CRC Press
6000 Broken Sound Parkway NW, Suite 300, Boca Raton, FL 33487-2742

and by CRC Press
4 Park Square, Milton Park, Abingdon, Oxon, OX14 4RN

CRC Press is an imprint of Taylor & Francis Group, LLC

ISBN: 978-0-367-50789-3 (hbk)
ISBN: 978-1-032-29076-8 (pbk)
ISBN: 978-1-003-05125-1 (ebk)

DOI: 10.1201/9781003051251

Typeset in LM Roman
by KnowledgeWorks Global Ltd.

Publisher's note: This book has been prepared from camera-ready copy provided by the authors

To Carrie, Trent, and Paige for their constant support and patience while Dad worked late nights on this book.

Contents

List of Figures xiii

List of Tables xv

Preface xix

Author xxiii

1 Introduction 1
- 1.1 What is Operations Research . . . 1
- 1.2 Purpose of this Book . . . 1
- 1.3 Range of Operations Research Techniques . . . 2
- 1.4 Relationship between Operations Research and Analytics . . . 3
- 1.5 Importance of Optimization . . . 3
- 1.6 Why R? . . . 4
- 1.7 Conventions Used in this Book . . . 6

2 Introduction to Linear Programming 7
- 2.1 What Is Linear Programming . . . 7
- 2.2 Two Variable Base Case . . . 7
- 2.3 Graphically Solving a Linear Program . . . 9
- 2.4 Implementing and Solving with `ompr` . . . 14
  - 2.4.1 Preparing to Implement the Linear Program . . . 14
  - 2.4.2 Implementing the Base Case with Piping . . . 20
- 2.5 Adding a Third Product (Variable) . . . 22
  - 2.5.1 Three Variable Base Case Formulation . . . 22
  - 2.5.2 Three Variable Base Case Implementation . . . 23
  - 2.5.3 Three Variable Case Results and Interpretation . . . 24
- 2.6 Linear Programming Special Cases . . . 24
  - 2.6.1 Case 1: No Feasible Solution . . . 25
  - 2.6.2 Case 2: Multiple Optima . . . 26
  - 2.6.3 Case 3: Redundant Constraint . . . 30
  - 2.6.4 Case 4: Unbounded Solution . . . 31
- 2.7 Abstracting the Production Planning Model . . . 33
- 2.8 Methods of Solving Linear Programs . . . 34
- 2.9 Exercises . . . 35

3 More Linear Programming Models 39
3.1 Types of LP Models . . . 39
3.2 The Algebraic Model . . . 39
3.2.1 Tips and Conventions for Algebraic Models . . . 40
3.2.2 Building the Generalized Model in R . . . 41
3.2.3 Examining the Results . . . 44
3.2.4 Changing the Model . . . 46
3.3 Common Linear Programming Applications . . . 47
3.3.1 Blending Problems . . . 47
3.4 Allocation Models . . . 50
3.4.1 Covering Models . . . 51
3.4.2 Transportation Models . . . 53
3.4.3 Transshipment Models . . . 57
3.4.4 Production and Inventory Planning . . . 59
3.4.5 Standard Form . . . 60
3.5 Vector and Matrix Forms of LPs . . . 63
3.6 Exercises . . . 67

4 Sensitivity Analysis 71
4.1 Base Case . . . 71
4.2 Shadow Prices . . . 72
4.2.1 Extraction and Interpretation . . . 72
4.2.2 Example of Adding an Hour to Assembly . . . 74
4.2.3 Shadow Prices of Underutilized Resources . . . 75
4.3 Reduced Costs of Variables . . . 76
4.3.1 Reduced Cost of Ants . . . 77
4.3.2 Reduced Price of Bats . . . 79
4.4 Using Sensitivity Analysis to Evaluate a New Product . . . 81
4.5 Exercises . . . 82

5 Data Envelopment Analysis 85
5.1 Introduction . . . 85
5.2 Creating the Data . . . 86
5.3 Graphical Analysis . . . 88
5.4 The Linear Programs for DEA . . . 90
5.4.1 An Explicit Linear Program for DEA . . . 90
5.4.2 A Generalized Linear Program for DEA . . . 91
5.5 Creating the LP – The Algebraic Approach . . . 93
5.6 Returns to Scale . . . 100
5.7 Multiple Inputs and Multiple Outputs . . . 105
5.8 Extracting Multiplier Weights from Sensitivity Analysis . . . 113
5.9 Slack Maximization . . . 116
5.10 DEA Packages . . . 119
5.11 DEA Model Building . . . 119
5.11.1 Selection of Inputs and Outputs . . . 120

5.11.2 Model Choices . . . 121
5.11.3 Application Area Expertise . . . 121
5.12 Further Reading . . . 122
5.13 Exercises . . . 122

6 Mixed Integer Optimization 125
6.1 Example of Minor Integrality Impact . . . 125
6.2 Example of Major Integality Impact . . . 128
6.3 The Branch and Bound Algorithm . . . 130
6.3.1 The LP Relaxation . . . 131
6.3.2 Subproblem I . . . 132
6.3.3 Subproblem III . . . 133
6.3.4 Subproblem IV . . . 135
6.3.5 Subproblem V . . . 136
6.3.6 Subproblem VI . . . 136
6.3.7 Subproblem VII . . . 138
6.3.8 Subproblem VIII . . . 139
6.3.9 Subproblem II . . . 139
6.4 Computational Complexity . . . 141
6.4.1 Full Enumeration . . . 142
6.5 Binary Variables and Logical Relations . . . 143
6.6 Fixed Charge Models . . . 146
6.6.1 Fixed Charge Example-Introduction . . . 147
6.6.2 Linking Constraints with "Big M" . . . 151
6.6.3 Fixed Charge Implementation . . . 154
6.7 Model Results and Interpretation . . . 155

7 More Integer Programming Models 157
7.1 Overview . . . 157
7.2 Revisiting the Warehouse Location Problem . . . 157
7.2.1 Implementing the Warehouse Model . . . 159
7.2.2 Solving the Warehouse Location Problem . . . 165
7.2.3 Warehouse Discussion . . . 172
7.3 Solving MIPs with Different Solvers . . . 173
7.3.1 Performance of `glpk` . . . 173
7.3.2 Performance of `symphony` . . . 174
7.3.3 Performance of `lpsolve` . . . 177
7.3.4 Performance of `gurobi` . . . 178
7.3.5 Comparing Results across Solvers . . . 179
7.3.6 Popularity of LP Solvers . . . 181
7.4 Solving Sudoku Puzzles using Optimization . . . 184
7.4.1 Introduction to Sudoku and Optimization . . . 184
7.4.2 Formulating the Sudoku Problem . . . 185
7.4.3 Implementing Sudoku in ompr . . . 187
7.4.4 Sudoku Discussion . . . 192

7.5 Exercises 192
7.6 Production Planning over Time 195
7.6.1 Implementing the Model 199
7.7 Additional Exercises 203

8 Goal Programming 205
8.1 Introduction 205
8.2 Preemptive Goal Programming 205
8.3 Policies for Houselessness 207
8.4 Mass Mailings 213
8.4.1 Formulating the State Mailing Model 214
8.4.2 Implementing the State Mailing Model 215
8.4.3 Frontloading the Work 219
8.5 Exercises 222

A A Very Brief Introduction to R 225
A.1 Purpose 225
A.2 Getting Started with R 225
A.3 Exercises 233

B Introduction to Math Notation 235
B.1 Purpose 235
B.2 Basic Summation Notation 235
B.3 Using LaTeX in RMarkdown 237
B.4 Inline Notation 237
B.5 Sums 238
B.6 Delimiters 238
B.7 Summary of Mathematical Notations 239
B.8 Sequences and Summation Notation 240
B.9 Applications of Summation 240
B.10 Double Summation 243
B.11 Applications of Double Summation 244
B.12 Exercises 244

C Troubleshooting 247
C.1 Overview 247
C.2 Model Building 247
C.2.1 Define and Formulate before Implementing 247
C.2.2 Failing to Look for Past Optimization Models 248
C.2.3 Misrendering of PDF 248
C.2.4 Blank Lines in LaTeX 249
C.2.5 Problems with PDF Creation 250
C.3 Implementation Troubleshooting 252
C.3.1 Errors in a Piped Model 252

C.3.2 Undefined Object in `ompr` 254
C.3.3 Unexpected Symbol in `ompr` 255
C.3.4 Name Conflicts between R and `ompr` 256
C.3.5 Blindly Reusing Code 260
C.4 General Debugging Tips 260
C.5 Getting Help 261

D Making Good Tables **263**
D.1 Importance of Tables in Modeling 263
D.2 Kable vs. Kbl 264
D.3 Table Footnotes with Kable 264
D.4 Setting Row and Column Names in Kable 265
D.5 Booktabs vs. Default 268
D.6 Using LaTeX in Kable Column Names 269
D.7 Fitting Tables to Page Width 270

Bibliography **271**

Index **273**

List of Figures

2.1 Drawing the First Constraint. 10
2.2 Adding the Assembly Constraint. 10
2.3 Adding the Testing Constraint. 11
2.4 Adding the Sensors Constraint. 11
2.5 Feasible Region for Drone Manufacturing. 12
2.6 Candidate Solutions. 12
2.7 Parallel Equal Profit Lines. 13
2.8 Optimal Solution. 13

3.1 Transportation Example. 54
3.2 Transshipment Application. 58
3.3 Relationship between ompr and solvers. 63
3.4 Relationship between ompr and Rglpk. 65

5.1 A Simple DEA Model for Store Management. 87
5.2 Store Benchmarking Example. 89
5.3 BCC (VRS) Efficiency Frontier. 102
5.4 Increasing (Non-Decreasing) Returns to Scale Efficiency Frontier. 104
5.5 Decreasing(Non-Increasing) Returns to Scale Efficiency Frontier. 105

6.1 LP Region and 12 IP Feasible Solutions. 129
6.2 LP Relaxation. 132
6.3 Subproblem I. 134
6.4 Subproblem III – An IP Feasible Solution. 135
6.5 Subproblem IV. 136
6.6 Subproblems V and VI. 137
6.7 Subproblem VII and VIII. 138
6.8 Subproblem VII Solution. 139
6.9 Subproblems VI and VII. 140
6.10 Branch and Bound Tree. 142

7.1 Map of Customer and Warehouse Locations. 164
7.2 Optional Warehouse Locations Customer Assignments. 170
7.3 LP Region and 10 IP Feasible Solutions. 171
7.4 Monthly Downloads of Optimzation Packages. 182

7.5 Monthly Downloads of Major ROI LP Plugins. 182
7.6 Monthly Downloads of ompr. 183

8.1 Map of Mailing Results. 220

C.1 Preview Ignores Blank LaTeX Line. 249
C.2 Rendering Error Caused by Blank LaTeX Line. 250
C.3 Name Conflict Error between R and `ompr`. 252
C.4 Inability to Focus on Error Source in a Piped Model. 253
C.5 Piped vs. Unpiped Model. 253
C.6 Finding an Error in an Unpiped Model. 253
C.7 Error from an Defined Variable. 255
C.8 Unexpected Symbol in ompr. 256
C.9 Error Due to Name Conflict between R and `ompr`. 258
C.10 Error as Displayed in Console from Run All Chunks. 258
C.11 Sweeping Out the Environment. 259

List of Tables

2.1 Two Variable Base Case 7
2.2 Base Case Two Variable Solution 18
2.3 Elements of the Constraints 20
2.4 Three Variable Base Case 23
2.5 First Solution of Multiple Optima Case 27
2.6 An Alternate Optimal Solution 28
2.7 Examples of Alternate Optimal Solutions 29
2.8 Data for Dog Drone Exercise 35
2.9 Characteristics with Painting Changes 36
2.10 Data for Rose City Roasters 36

3.1 Profit per Product 41
3.2 Resources Used by Each Product and Amount Available 42
3.3 Optimal Production Plan 44
3.4 Resources Used in Optimal Solution 46
3.5 Revised Optimal Production Plan 47
3.6 Production Plan for Base Case 49
3.7 Compare Baseline and Production Plan with Blending Constraint 50
3.8 Trevor Trail Mix Company Ingredients 51
3.9 Data for Trevor Trail Mix Company 52
3.10 Demand and Supply Constrained Transportation Problems . . 55
3.11 Costs for Transportation Application 56
3.12 Optimal Transportation Plan 57
3.13 Solution to Standard Form 63
3.14 Transportation Cost between Cities 67
3.15 Staff Scheduling 70

4.1 Base Case Production Plan 72
4.2 Shadow Prices of Constrained Resources 73
4.3 Production Plan with One Additional Assembly Hour 75
4.4 Production Plan with 10,000 More Sensors 76
4.5 Reduced Costs of Variables 77
4.6 Resources Used by an Ant and Shadow Prices of Resources . . 78
4.7 Resources Used by a Cat and Their Shadow Prices 79
4.8 Resources Used by a Bat and Their Shadow Prices 79

4.9 Impact of a Forced Change in Bats 81
4.10 Resources Used by a Dog and Their Shadow Prices 82
4.11 Adding Eels 82

5.1 First Dataset for DEA 88
5.2 Input-Oriented Envelopment Analysis for DMU B 96
5.3 Input-Oriented Envelopment Analysis for DMU A 97
5.4 Input-Oriented Efficiency Results 99
5.5 Results with Inefficient DMU Columns Removed 99
5.6 Input-Oriented VRS Envelopment Results 101
5.7 Returns to Scale Envelopment Constraints 103
5.8 Input-Oriented Model with Decreasing Returns to Scale 104
5.9 Input-Oriented Increasing Returns to Scale Model Results . . . 104
5.10 Comparison of Efficiency under Alternate Returns to Scale Assumptions 105
5.11 Simplified Data for University Technology Transfer 106
5.12 Selected University Technology Transfer Characteristics 108
5.13 Results from University Technology Transfer Example (CRS) . 110
5.14 Comparison of CRS vs. VRS Efficiency Scores 112
5.15 Input and Output Weights for Last DMU 114
5.16 Multiplier Weights from UTT Example 115
5.17 Camera Data 123

6.1 Production Plan with Continuous Variables 126
6.2 Production Plans Based on Variable Type 128
6.3 Production Plan-Continuous vs. Integer 130
6.4 Acme's Production Plan Based on the LP Relaxation 132
6.5 Production Plan Based on Subproblem I 133
6.6 Integer Valued Production Plan Based on Subproblem III . . . 134
6.7 Production Plan Based on Subproblem IV 135
6.8 Production Plan Based on Subproblem V 136
6.9 Infeasible Production Plan Based on Subproblem VI 137
6.10 Integer Valued Production Plan Based on Subproblem VII . . . 138
6.11 Infeasible Production Plan Based on Subproblem VIII 139
6.12 Production Plan Based on LP Subproblem II 140
6.13 Acme's Integer Optimal Production Plan Based on Subproblem VII 141
6.14 Branch and Bound's Sequence of LPs 141
6.15 Data for the Outtel Example 144
6.16 Relationships between Projects A, B, and the Need for Chip Architects 145
6.17 Widget Characteristics 147
6.18 Widget Production Requirements 147
6.19 Base Case Solution for Fixed Charge Problem 150
6.20 Truth Table for Widget 2 Variables 150

6.21 Fixed Charge Problem 154
6.22 Manufacturing Resource Usage 156

7.1 Locations of First Six Customers 163
7.2 Raw Results for Vx Variable 167
7.3 Subset of Customers (i) Assigned to Warehouses (j) 168
7.4 XY Coordinates to Draw Customer-Warehouse Routes 168
7.5 Count of Customers Assigned to Each Operating Warehouse . . 169
7.6 Summary of Results for Warehouses Used 169
7.7 Comparison of Results from Warehouse Customer Assignment Across Solvers 180
7.8 Product Demand Over Time 196
7.9 Production Planning Over Time 202
7.10 Production Planning 203

8.1 Policy Options for Addressing Houselessness 207
8.2 Group 1's Ideal Solution 209
8.3 Comparing Homelessness Solutions 210
8.4 Solution by Minimizing Sum of Deviations 210
8.5 Solution by Minimizing Sum of Percentage Deviations 211
8.6 Minimax Solution 213
8.7 Number of Customers for First Six States 214
8.8 Example of Some States Assigned to Week 1 218
8.9 Display of States by Week 219
8.10 Solving Progress for Modified Mailing Model 221

A.1 Example Using kable with booktabs 229
A.2 Scalar Multiplication of Matrix b to Make Matrix c 229
A.3 Transposition of Matrix b 230
A.4 Original Matrix b 231
A.5 Second Row of b 231
A.6 Third Column of b 231
A.7 Last Element of b 231
A.8 Combined Matrix 232
A.9 Combined Matrix with Explanation of Source 232
A.10 Column Binding of Matrices b and c 232
A.11 Row Binding of Matrices b and c 233

B.1 Commonly Used Optimization Modeling Notations 239
B.2 Production Costs for Product 1 241
B.3 Itemized Production Costs for Three Products 242

D.1 Footnotes in Tables Using kbl 265
D.2 Retain Row Names by Using 'row.name=T' 266
D.3 Drop Row Name by Using 'row.name=F' 267
D.4 Adding Row Names as a Leading Column 267

D.5 Replacing Row Names in Matrix Before kable 267
D.6 Default Format Using kable 268
D.7 Kable Using booktabs 269
D.8 Using LaTeX in Row and Column Names 269
D.9 Scaling Table to Fit Page 270

Preface

This book covers using R for doing optimization, a key area of operations research, which has been applied to virtually every industry. The focus is on linear and mixed integer optimization.

Pedagogically, since the late 1990s, I had been a proponent of teaching introductory optimization using spreadsheets as a way to bridge the barrier between data and applications as well as teaching better spreadsheet practices. While this worked well, the disconnect with algebraic modeling was always a problem. Excel add-ins enabling algebraic tools were rich but not seamless. In 2018, I decided to switch to R for a variety of reasons:

- With the rapid rise of interest in data analytics, it became important to introduce a data analytics platform.
- Working technology and business professionals should have an awareness of the tools and language of data analysis to deal with consultants, data scientists, and others.
- New software and extensions reduced the barrier to entry for people to start using data analytics tools.

Philosophically, the book emphasizes creating formulations before going into implementation. This is important for the following reasons:

- Reading the literature is based on the algebraic representation.
- An algebraic representation allows for clear understanding and generalization of large applications.
- Writing formulations is necessary to explain and convey the modeling decisions made.
- Becoming proficient in the mathematical language used will be transferable to other analytical models.
- Dividing the modeling stage from the implementation enables better debugging of models avoids what is sometimes referred to as spaghetti code in programming.
- Separating enables the reader to readily transition to another language for implementation in the future (ex. Python, Julia, AMPL, GAMS) with what will seem like little more than a dialect change.

Intended Audience

This book is written for people with at least a passing familiarity with R or another programming environment, but Appendix A helps jumpstart the motivated reader to jump in without a background. Some familiarity with mathematics is used throughout the book at the level of subscripts and summations, but refreshers are provided in Appendix B. It is assumed that the reader is willing to use R and get their hands dirty with experimenting.

Features of this Book

- Providing and explaining code, sometimes repeated in different places so that examples are relatively clear and self-contained.
- An emphasis on creating algebraic formulations before implementing.
- A focus on application rather than algorithmic details.
- Embodying the philosophy of reproducible research – the book is regularly rebuilt with all analyses automatically rerun and most tables and figures rebuilt.
- Making use of open source tools for analysis to ensure the readers have permanent access to powerful optimization tools.
- Contributing to the open source community – all materials are available on the author's github repository.
- Demonstrating common debugging practices with a troubleshooting emphasis specific to optimization modeling using R.
- Providing code chunks liberally that readers can adapt to their own applications.

Instructor Notes

This book has been used multiple times for a ten-week graduate course on operations research emphasizing optimization. It can be used for graduate and undergraduate courses for people without a background in optimization and varying levels of mathematical backgrounds. The focus is on applications (formulating, implementing, and interpreting rather than algorithms. The book could be used as a supplement in a more theoretical or algorithm-oriented class.

Acknowldegments

I would like to thank many people for their contributions, collaborations, and assistance over the years. All errors are my fault, though.

- **Dirk Schumacher,** author of the `ompr` package used throughout this book
- **Dr. Dong-Joon Lim,** applications and methodological work in DEA
- **Dr. Gerry Williams,** application of DEA to construction contracting
- **Dr. Janice Forrester,** application of DEA to the energy industry
- **Dr. Scott Leavengood,** application of DEA to wood products
- **Dr. Oliver (Lane) Inman,** early work on TFDEA
- **Dr. Maoloud Dabab,** many suggestions over time
- **Dr. K. Louis Luangkesorn,** author of the first vignette on using `glpk` in R
- **Dr. Chester Ismay,** contributions to the Portland and broader R community
- **Dr. Jili Hu,** rich interactions during his sabbatical in Portland
- **Dr. Nina Chaichi,** many suggestions over the years
- **Tom Shott,** primary author of the `TFDEA` package
- **Aurobindh Kalathil Kumar,** PhD student, many suggestions over time
- **Kevin van Blommestein,** earlier DEA & R work
- **William (Ike) Eisenhauer,** LaTeX formulation improvements
- **Andey Nunes,** coding improvements
- **Christopher Davis,** graphical example of LP
- **Thanh Thuy Nguyen,** fixed charge example
- **Roland Richards,** formatting assistance
- **Caroline Blackledge,** contributed to appendix
- **Alexander Keller,** contributed to appendix
- **Shahram Khorasanizadeh,** contributed to appendix
- **Jose Banos,** formatting assistance
- **Jon Syverson,** frequent and thoughtful feedback on early drafts
- **Dawei Zhang,** further editorial work on the book
- **Navdeep Singh,** assistance with formatting and additional exercises
- **Ketsaraporn Kaewkhiaolueang,** assistance with proofreading

In addition, several groups have been of tremendous help:

- The Portland Meetup, R User's Group
- The Extreme Technology Analytics Research Group
- Past ETM 540/640 Operations Research classes and, in particular, the 200 students since 2019 that have used drafts of this book.
- My advisors and mentors from Georgia Tech, Dr. Gunter Sharp and Dr. Ronald Bohlander.

Most of all, I would like to also express my appreciation for my family's patience while working on this book with many late nights: Carrie, Trent, and Paige.

Tim Anderson 2021-12-22

Author

Dr. Timothy Anderson is the Department Chair of Engineering and Technology Management at Portland State University. He earned an Electrical Engineering degree from the University of Minnesota, as well as both M.S. and Ph.D. degrees in Industrial Engineering from the Georgia Institute of Technology. Dr. Anderson has been the Program Chair or Co-Chair twenty times for PICMET, the Portland International Conference on the Management of Engineering and Technology since 1997. He leads the Extreme Technology Analytics research group. He was President of Omega Rho, the International Honor Society for Operation Research. With over 60 refereed publications, current research interests include analytics, benchmarking, technology forecasting, data mining, and new product development. He helped lead the creation of six graduate certificates at Portland State, including Business Intelligence and Analytics.

Dr. Anderson is an enrolled member of the Pokagon Band of Potawatomi, making him one of the few engineering department chairs in the country who is also a member of a Native American tribe. He values diversity and works to ensure that future engineers and scientists better represent the full community. He has been recognized as a Sequoyah Fellow of the American Indian Science and Engineering Society.

1

Introduction

1.1 What is Operations Research

Operations Research is the application of mathematics and scientific approaches to decision making. It is a field deeply connected with applications.

The field of Operations Research has strong roots in World War II as nations and militaries urgently tried to best make use of their resources and capabilities. Leaders asked people from various disciplines to help improve a variety of problems such as submarine hunting, mining of resources, and production planning. Essentially, these tools are scientific approaches for making decisions for the design and operation of a system with limited resources. The tools that they used came from combining applied math and scientific approaches resulting in tremendous improvements.

After the war, many of the same approaches were also then used for business applications. Military groups most commonly used the term Operations Research while business groups often used the term Management Science.

Neither term, Operations Research nor Management Science is perfect. Operations Research is often confused with the strongly overlapping area of Operations Management, and also, the word research implies to some that it is strictly theoretical. Conversely, Management Science might imply that the rest of management is non-scientific or that it is just limited to business applications. In this book, we will use the term Operations Research to mean both Operations Research and Management Science as well as often abbreviate it as OR.

1.2 Purpose of this Book

There are many comprehensive introductions to Operations Research available (Hillier and Lieberman, 2020) (Winston, 2003), and others. In the past, I had

DOI: 10.1201/9781003051251-1

emphasized a spreadsheet-centric based introduction to operations research such as (Ragsdale, 2017) and (Baker, 2015), which had the side benefit of also encouraging better spreadsheet usage practices. Unfortunately, the spreadsheet metaphor constrains the reader's perspective, often leading people to tie models to closely to the data analyzed, specific dimensions of a particular problem, and the rows/columns construct often prevents people from thinking of richer problems. A more rigorous and tool-agnostic coverage of optimization can be beneficial, for example, Tovey's recent book stands out (Tovey, 2021). The goal of this book is to provide a timely introduction to Operations Research using a powerful open-source tool, R. It was written to support the ETM 540/640 Operations Research course at Portland State University, which is taught on a quarter basis for people without a background in Operations Research or R. As such the current scope is limited to what can be readily accomplished in a 10 week quarter, giving an introduction to the tools and methods available. There are many other Operations Research tools available besides R.

This book is meant to be a hybrid, both serving as an introduction to R and to Operations Research in the context of optimization.

You can always access the current version of this book on Github. `https://github.com/prof-anderson/OR_Using_R`

1.3 Range of Operations Research Techniques

Operations Research covers many different techniques. They can be classified in a wide range of ways. Some of the more common approaches are:

- Optimization
- Simulation
- Queuing Theory
- Markov Chains
- Inventory Control
- Forecasting
- Game Theory
- Decision Theory

Each of these topics can require a full book on its own. The focus of this book is on the most widely used operations research technique, optimization, and more specifically, linear and mixed-integer optimization.

In particular, Optimization and Simulation can be further subdivided into separate sub-areas that represent entire specialties of their own. In this book, we currently limit ourselves to the field of Optimization, but Chapter 7 incorporates elements of simulation. The interested reader is welcome to explore each area further.

Analysis methods are typically characterized as either descriptive or prescriptive {Prescriptive models}. A descriptive technique would be one that describes the current state of the system. A prescriptive technique is one that would prescribe a certain course of action to achieve the desired outcome. For example, weather forecasting is a very sophisticated descriptive discipline since we don't generally have control over short-term weather. On the other hand, if we were to look at intervention approaches such as cloud seeding, we might try to come up with a prescriptive application. Prescriptive applications are abundant and can often be wherever decisions need to be made.

1.4 Relationship between Operations Research and Analytics

The field of Operations Research significantly predates that of the term analytics, but they are closely linked. This is strongly demonstrated by the leading professional organization for operations research, INFORMS (The Institute for Operations Research and Management Science), developing a professional certification for analytics. This certification process had leadership from the industry representing the diverse users of Operations Research techniques.

1.5 Importance of Optimization

Optimization has fundamentally changed the world in diverse applications.

To highlight this, let's look at how a broad range of industries affected by Covid-19 have been using optimization.

Supply Chain Management: Amazon executives have given keynote speeches at INFORMS international conferences discussing how optimization is used in designing their network of warehouses and delivery services. The rapid growth of online shopping and acceleration of ever-faster delivery options in the last decade relies on optimization, even more so as supply chains have been disrupted by Covid-19 and product demand for certain items has skyrocketed.

Scheduling of sports events: Major League Baseball retained optimization experts for scheduling all regular season games. Dr. Michael Trick from Carnegie Mellon University recounted this his work in an INFORMS Omega Rho 2017 Distinguished Lecture. The need for optimization in this domain has only increased with the additional complexity of the constant drumbeat of sporting event cancellations and rescheduling due to Covid-19. Analytics revolutionized sports in many ways as documented by the Michael Lewis book, Moneyball, the impact of optimization on scheduling may be just as significant.

Medical Staff Scheduling Management of human resources in the health care area can be a matter of life and death. Having fully staffed emergency departments is a critical issue that is now getting featured regularly on the news due to Covid-19, but the need for optimization was recognized long before the pandemic. Scheduling is so complicated of nursing with that it has created a vertical market industry of companies that tools with built-in optimization engines to provide 24x7 coverage across departments and critical skills.

Transportation Airlines often have their own optimization teams in-house to continually reexamine their network of flights, planes used, and crews. In addition to the steady-state, using optimization models to rapidly adjust to disruptions, whether it is caused by weather, Covid-19, security, or maintenance issues.

In this book, we provide tools for engaging with these kinds of applications.

Optimization provides a framework for analyzing a broad range of

1.6 Why R?

This book adopts the platform of R for doing optimization. R is a popular, open-source programming language developed by statisticians, originally as an open-source alternative to the statistics language S. As such, it has very strong numerical methods implementations. Given the open-source nature, it has been popular for researchers, and many of these researchers have written their own packages to extend the functionality of R. These packages cover a tremendously broad range of topics, and well over 18,000 are available from CRAN (the Comprehensive R Archive Network) as of December 2021.

In about 2012, my Extreme Technology Analytics Research Group was facing a decision. We had been using proprietary analytics tools for years but had problems with software licensing and the limited range of specialized languages.

We wanted a single platform that could do both statistics and optimization, was freely available with no ongoing license renewal and was readily extensible.

In the end, we debated between Python and R. While R grew out from statisticians to encompass broader areas, Python grew from a general purpose programming language to add features to address a broad range of different application areas including statistics. While they both have deep roots with over 20 years of active development, this also means that aspects of their original designs are somewhat dated and carry forward the burden of legacies of past decisions when computers were more limited, and the meaning of a big dataset would have been much smaller.

Python had some advantages from the optimization side, while R had advantages from the statistical side. In the end, we chose R because Python's numerical methods capabilities were not quite as rich. Since that time, both R and Python have significantly matured.

As time has gone by, R has developed robust toolsets such as RStudio, making it more powerful and easier to use. R's extensive package community has become so diverse and powerful that proprietary statistics software such as SPSS and JMP now promote that they can use R packages. The result is that R is still a great choice for analysts and code developed for R should be usable for a long time.

As should be expected, new tools continue to arise. In particular, Julia is a new, modern language developed around numerical analysis. Industry adoption of Julia is far behind R and Python given their head start of multiple decades. Even if Julia or some other environment does someday become more popular, skills developed using R will be readily transferable and provide an excellent foundation for learning new languages.

A PhD student that I knew was preparing for his final defense and did a practice presentation with his advisor and a professor. He was proud of a well-polished 250-page thesis with dozens of tables of numerical results. The analysis included dozens of regressions, all with similar form and dependent variables. The professor asked him why his dependent variable was the number of projects rather than projects per year. The student thought for a moment and realized that all of his analysis needed to be redone. This would take days or weeks if he had used a GUI-based tool. Using R and a scripting approach, he did a search and replace for the dependent variable, cross-checked results, and met the next day to update that professor and his advisor of the new (and much better!) results. He successfully defended his thesis a couple of weeks later.

1.7 Conventions Used in this Book

I've adopted the following conventions in this book.

- R code fragments, including packages, will be denoted using monospaced text such as adding two R data objects `a+b`.
- Mathematical symbols and equations will be italicized. For example, if two variables are mathematical symbols and are being added, then it would be $a+b$.
- Tables of information and code fragments are provided liberally and at times erring on the side of being repetitious to assist in the interpretation and reproducing analyses.
- There may be more efficient ways of performing operations, but the focus is on readability.
- When formatting piping operators such as `|>`, an intent is made to line them up for readability.

2

Introduction to Linear Programming

2.1 What Is Linear Programming

Linear programming is a tool for optimization. It is a widely used tool for planning, scheduling, resource allocation, and many other applications.

2.2 Two Variable Base Case

We will use a recurring example throughout the next couple of chapters of being a small specialty drone manufacturer making animal inspired exploring drones. These drones can be used in search and rescue operations or other functions where a wheeled vehicle may not suffice. Each of the products is named after an animal that describes its general design and function. The first two types of drones are named the Ant and the Bat. The Ant is a small but precise drone, while the Bat is a flying drone.

The goal is to make the most profitable mix of drones.

Each drone requires time in machining, assembly, and testing and sensors. For example, an Ant requires one hour of machining time, three hours of assembly, and two hours of testing. It uses two sensors, and the net profit is \$10. The characteristics of Bats are similar but different and are shown in the following table.

TABLE 2.1 Two Variable Base Case

Characteristic	Ants	Bats	Available
Profit	\$7	\$12	
Machining	1	4	800
Assembly	3	6	900
Testing	2	2	480
Sensors	2	10	1200

DOI: 10.1201/9781003051251-2

A simple LP now is to find the production plan of products that results in the most profit. In order to do so, we need to define certain key items:

- the goal(s)
- the decisions
- the limitations

Let's start with the goal(s). In this case, the production manager is simply trying to make as much profit as possible. While cost cutting is also a goal for many organizations, in this case and many applications, profit maximization is appropriate. Maximizing profit is the referred to as the *objective* of the model.

People new to linear programming will often think of the decisions as the amount of each resource to use. Instead, the decisions in this case would be how much to make of each particular product. This drives the resource usage, and the resource usage is a byproduct of these decisions. These decisions can take on a range of values and are, therefore, called *decision variables.*

The decision variables are then combined in some way to reflect the performance with respect to the organization's objective. The equation combining the decision variables to reflect this is then the *objective function.* For now, we will assume that there is a single objective function, but we will allow for multiple objectives in Chapter 8.

Lastly, what is limiting the organization from even better performance? There are typically many limits such as the number of customers, personnel, supplier capacity, etc. In this case, we focus on a set of resource limits based on staffing in different centers and the supply of sensors. Since these limitations constrain the possible values of decision variables, they are called *constraints.*

Every optimization model can be thought of a collection of:

- an objective function (goal)
- decision variable(s) (decisions)
- constraint(s) (limitations)

Let's put things together in the context of this application. In the base case, our objective function is to maximize profit. We can't express it, though, until we define our decision variables. It is good practice to very clearly and precisely define decision variables. While this case is very straightforward, the definition of variables can get much more complicated as we move into richer and larger models.

Let's define the variables:

- Ants = # of Ant drones to make
- Bats = # of Bat drones to make

Our objective function and constraints can now be written as the optimization model shown in the following formulation.

$$
\begin{aligned}
\text{Max } & 7 \cdot Ants + 12 \cdot Bats & \text{[Profit]} \\
\text{s.t.: } & 1 \cdot Ants + 4 \cdot Bats \leq 800 & \text{[Machining]} \\
& 3 \cdot Ants + 6 \cdot Bats \leq 900 & \text{[Assembly]} \\
& 2 \cdot Ants + 2 \cdot Bats \leq 480 & \text{[Testing]} \\
& 2 \cdot Ants + 10 \cdot Bats \leq 1200 & \text{[Sensors]} \\
& Ants,\ Bats \geq 0 & \text{[Non-negativity]}
\end{aligned}
$$

Note that since the objective function and each constraint is a simple linear function of the decision variables, this is what we call a *linear* programming model or LP for short. It would not be linear if any nonlinear function is made of the decision variables. For example, squaring a decision variable, using conditional logic based on the variable value, multiplying two variables, or dividing a variable by a function of a variable. These and other issues would then require using nonlinear programming or NLP. NLP is also widely used but has limitations.

Linear programming: (LP, also called linear optimization) is a method to achieve the best outcome (such as maximum profit or lowest cost) in a mathematical model whose requirements are represented by linear relationships.

It is impressive the number of situations that can be modeled well using linear programming. Keeping to the world of linear programming, in general, allows for finding the very best solution to very big problems in a short amount of time. For example, it is common for practitioners to be analyzing real-world problems with hundreds of thousands of decision variables and constraints.

2.3 Graphically Solving a Linear Program

Given that the linear program has two variables, it can be represented in two-dimensions making it straightforward to draw and visualize. Before we analyze it using R, we will walk through a graphical representation.

We do this by iteratively adding each constraint and trimming the feasible region. The line can be drawn by looking at the constraint from two perspectives. First, as if only Ants are produced and second as if only Bats are produced. This gives specific point on the horizontal and vertical axes, respectively, which are simply connected.

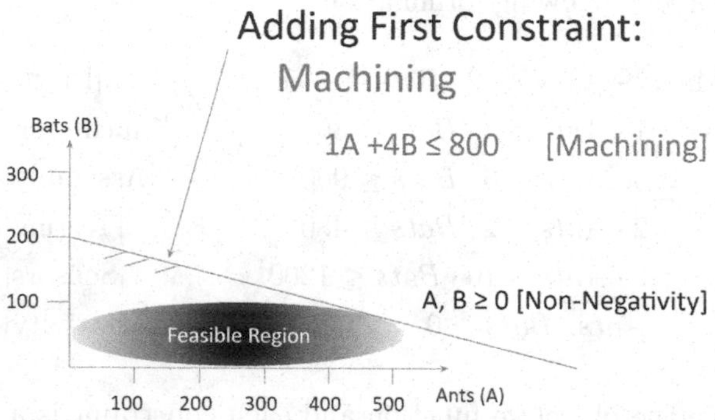

FIGURE 2.1 Drawing the First Constraint.

The constraints are differentiated by color, and the diagonal lines of the same color drawn off the constraint are a hatching line indicating the direction of the inequality constraint. In this case, each of the constraints are less than or equal to constraints and are indicating that this constraint includes the line and the region under the line. If it had been a greater than or equal to constraint, the hatching would have been rising above the constraint line. If it had been an exact equality constraint rather than an inequality constraint, it would have been just the line with no hatching above or below.

In each figure, we illustrate the feasible region as a grayed oval, but it is the whole region.

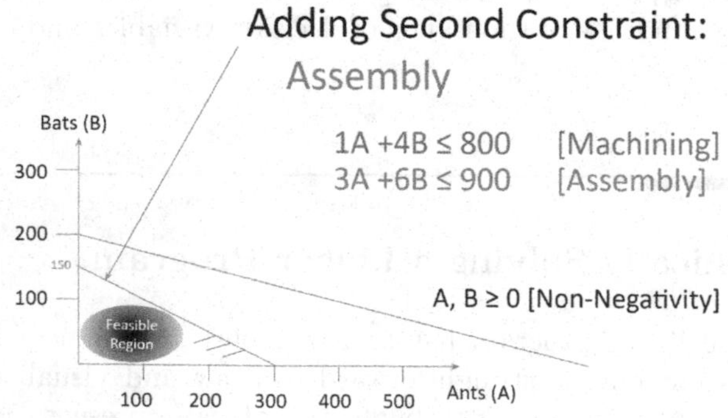

FIGURE 2.2 Adding the Assembly Constraint.

The assembly constraint is drawn as a red line connecting a production plan of just 300 Ants on the horizontal axis to just 150 Bats on the vertical axis.

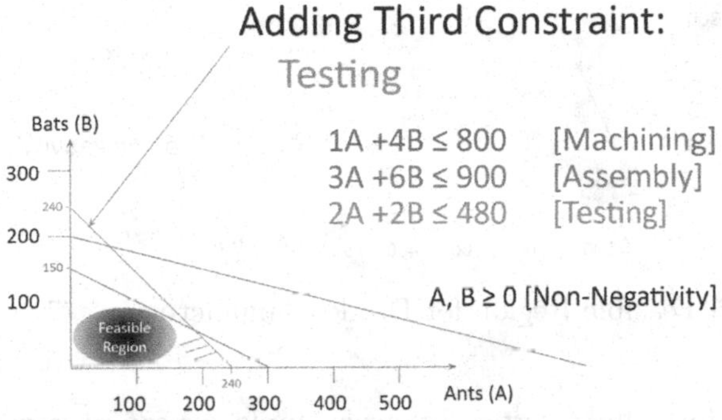

FIGURE 2.3 Adding the Testing Constraint.

Since both products require two hours of testing and there are 480 hours available, the Green Testing constraint connects producing just 240 Ants and just 240 Bats.

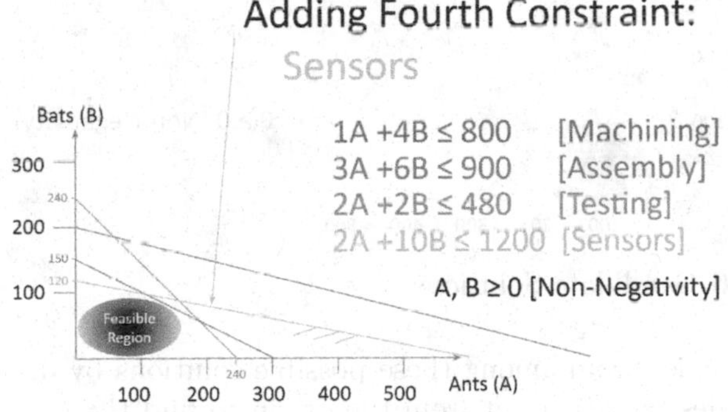

FIGURE 2.4 Adding the Sensors Constraint.

We now draw our fourth and final constraint – Sensors. This constraint is done in yellow, so it may be more difficult to see, but it goes from an Ants only production plan of 600 units to a bats only production plan of 120.

This now results in a feasible region that satisfies all of the constraints simultaneously. Every point in this region is feasible or possible in that it does not violate any of the constraints. There is an infinite number of possible points in this feasible region.

Given that we are doing linear programming and are trying to maximize or minimize a certain objective function, the optimal solution will be at a corner

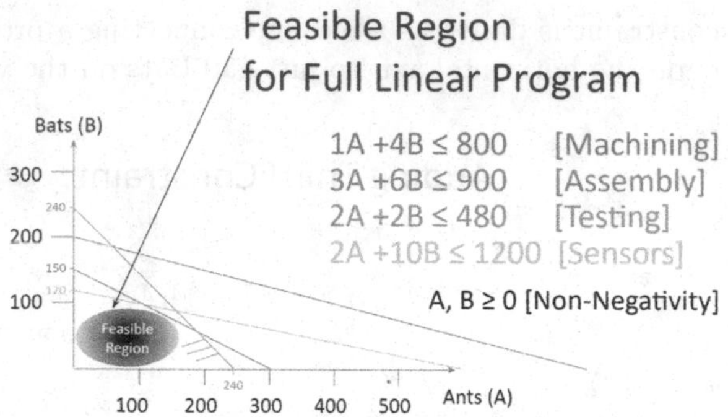

FIGURE 2.5 Feasible Region for Drone Manufacturing.

point or more formally, a vertex. In our example, we can narrow our attention from the full feasible region to just the five vertices or candidate solutions.

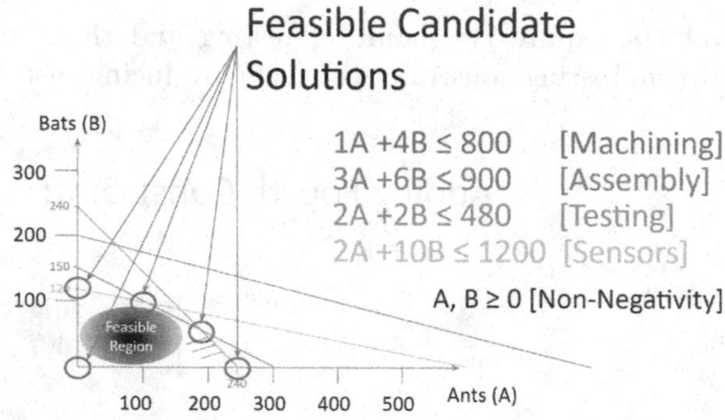

FIGURE 2.6 Candidate Solutions.

Lastly, we can pick from among these possible solutions by drawing parallel, equal profit lines radiating out from the origin to find the last line that still touches the feasible region. The figure to the right shows one line with a profit of \$1400. A second line is drawn that last touches the feasible region is at \$1980.

Graphically, we could read the number of Ants and the number of Bats to produce. Accuracy would be limited to the drawing skills and tools used.

Rather than relying on this graphical approach for determining the coordinates of each point drawn, each point in a two-dimensional space is defined by the intersection of two lines. In this case, the optimal solution (and all of the other candidate solutions too) can be defined by the intersection of two constraint lines, rewritten as exact equalities. These can be solved as a system of equations with two equations and two unknowns. The optimal solution here

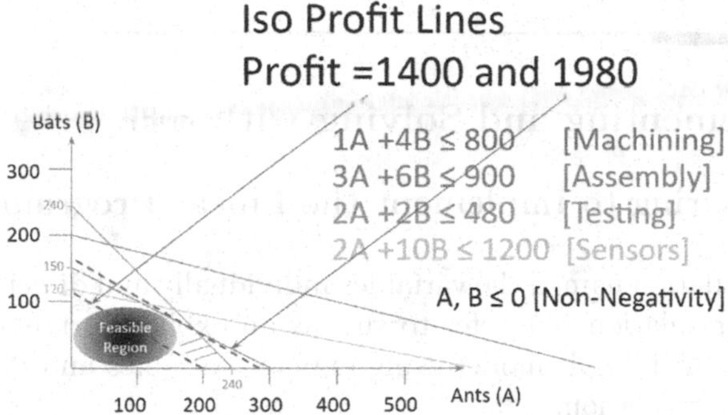

FIGURE 2.7 Parallel Equal Profit Lines.

is defined by the intersection of the Assembly (Red) and the Testing (Green) constraints. Just solve $3A + 6B = 900$ and $2A + 2B = 480$ for A and B.

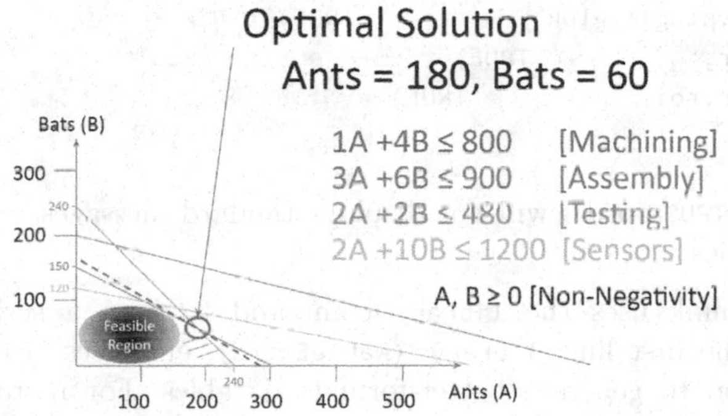

FIGURE 2.8 Optimal Solution.

This approach of graphical solving is helpful for understanding how optimization works but does not scale up to larger dimensions. Three variables corresponds to three dimensions and can, therefore, be drawn using various drawing programs but are harder to visualize on a printed page or computer screen.

Alas, real optimization problems typically have far more than three variables – perhaps tens of thousands or hundreds of thousands of variables. In general, we need to use computer tools for solving larger problems. In our case, we will adopt `R` and `ompr`, but there are a wide range of similar tools available.

2.4 Implementing and Solving with ompr

2.4.1 Preparing to Implement the Linear Program

Our first formulation named the variables individually and directly entered the data in the formulation. We refer to this as an explicit formulation. We will implement our first R LP model using explicit variables and data consistent with the first formulation.

First, let us load the required packages using the `library` command, and then we will move on to the actual implementation.

```
library (kableExtra, quietly = TRUE)
library (ROI, quietly = TRUE) # R Optimization Interface
library (ROI.plugin.glpk, quietly = TRUE) # Plugin for solving
library (ompr, quietly = TRUE) # Core ompr package
library (ompr.roi, quietly = TRUE) # Glue for ompr to use ROI
```

The `quietly=TRUE` option will not display standard messages returned when loading libraries.

This code chunk uses the `library` command to load a series of useful packages.[1] The first line, `library (kableExtra)`, enhances `knitr's` built-in `kable` function to generate better-formatted tables. For more details, see Appendix C. The following lines similarly load packages providing the optimization functions that we will be relying on frequently. The `ROI`package is the R Optimization Interface for connecting various optimization solvers to R. The `ROI.plugin.glpk` provides a connection between the `glpk`solver and R through ROI. While this would be sufficient, we are going to make use of the `ompr`package by Dirk Schumacher to provide algebraic representations for linear programs. The `ompr` package also requires a connector to ROI, aptly named `ompr.roi`.

Now we move on to implement and solve the linear program. Let's go through it step by step.

[1] As noted earlier, if these packages are not preinstalled, you may need to install them using the install.packages function or from the Packages tab in RStudio.

```
model0  <- MIPModel()      # Initialize an empty model
```

The first line initializes an empty model using the `MIPModel` command and stores it in `model0`. We could pick any valid R object name in place of `model0`, but we are using this to indicate that it is our initial model. The term MIP used in the function call is an acronym for Mixed Integer Program. This is a more general case of LP and will be discussed in greater detail later. We can see that the model is empty by simply displaying the summary of `model0`.

```
model0
```

```
## Mixed integer linear optimization problem
## Variables:
##   Continuous: 0
##   Integer: 0
##   Binary: 0
## No objective function.
## Constraints: 0
```

The summary states that there are no constraints or variables. Next, we will add variables.

```
model0a <- add_variable(model0, Ants,
           type = "continuous", lb = 0)
model0b <- add_variable(model0a, Bats,
           type = "continuous",lb = 0)
```

The first line takes `model0` and adds the `Ants` variable to it, creating an enhanced `model0a`. Note that the continuation of the first line specifies the type of variable and whether it has a bound. The `Ants` variable is continuous (as compared to integer or binary, which we will get to in chapter 6) and non-negative. The variable is made non-negative by setting a zero lowerbound (`lb=0`). The lowerbound can be set to other values such as a minimum production level of ten. Also, upperbounds for variables can be set using `ub` as a parameter.

The *Bats* variable is added in the same way to `model0a` creating `model0b`. Let's check the new model.[2]

[2]We are creating a series of models here as we keep adding one element at a time and appending a letter to `model0` to differentiate them. The result is that we will have many

```
model0b
```

```
## Mixed integer linear optimization problem
## Variables:
##   Continuous: 2
##   Integer: 0
##   Binary: 0
## No objective function.
## Constraints: 0
```

Next, we can add the objective function. We set the objective function as well as declaring it to be a *max* rather than a *min* function.

```
model0c<-set_objective(model0b,7*Ants+12*Bats,"max")
```

Now we move on to adding constraints.

```
model0d<-add_constraint(model0c, 1*Ants+ 4*Bats<= 800) # machining
model0e<-add_constraint(model0d, 3*Ants+ 6*Bats<= 900) # assembly
model0f<-add_constraint(model0e, 2*Ants+ 2*Bats<= 480) # testing
model0g<-add_constraint(model0f, 2*Ants+12*Bats<=1200) # sensors
```

Notice that we do not need to include non-negativity constraints since they were in the earlier definition of the variables as lower bounds of zero. Let's take a look at the summary of the model after all of the constraints have been added.

```
model0g
```

```
## Mixed integer linear optimization problem
## Variables:
##   Continuous: 2
##   Integer: 0
##   Binary: 0
```

incrementally enhanced versions of `model0`. Other than for demonstration purposes in our first model, there is no need to incrementally save each version of the file. We could replace each of the assignments by just resaving it to `model0`.

```
## Model sense: maximize
## Constraints: 4
```

Our model has the three core elements of an optimization model: variables, constraints, and an objective function. Let's go ahead and solve `model0g`. The LP can be solved using different LP engines – we'll use `glpk` for now. The results of the solved model will be assigned to `result0`. `glpk` is the R package name for the "GNU Linear Programming Kit", which is intended for solving large-scale linear programming (LP), mixed integer programming (MIP), and other related problems.

```
result0 <- solve_model(model0g, with_ROI(solver = "glpk"))
```

We built up the model line by line incrementally taking the previous model, adding a new element, and storing it in a new model. Note that we could keep placing them into the original model if we had wished such as the following. In this case, we take the previous command's output, `model0`, add an element to it, and then store it back in the same object, `model0`. We keep doing it for each element until we are done. This is equivalent to assigning the value of *2* to *a*. We then multiply *a* by *3* and assign the result again to *a*.

```
model0 <- MIPModel()          # Initialize an empty model
model0 <- add_variable(model0, Ants, type = "continuous", lb = 0)
model0 <- add_variable(model0, Bats, type = "continuous", lb = 0)
model0 <- set_objective(model0, 7*Ants + 12*Bats, "max")
model0 <- add_constraint(model0, 1*Ants + 4*Bats <= 800)    # machining
model0 <- add_constraint(model0, 3*Ants + 6*Bats <= 900)    # assembly
model0 <- add_constraint(model0, 2*Ants + 2*Bats <= 480)    # testing
model0 <- add_constraint(model0, 2*Ants + 10*Bats <= 1200) # sensors

result0 <- solve_model(model0, with_ROI(solver = "glpk"))

result0$solution
```

```
## Ants Bats
##  180   60
```

Let's use the `kable` function to display the results nicely.

```
res0 <- cbind(objective_value (result0),
          get_solution(result0, Ants),
          get_solution(result0, Bats))
colnames(res0)<-list("Profit","Ants","Bats")
rownames(res0)<-list("Solution")
```

Notice that we used two ways to get the solution of decision variables from `ompr`. In the first case, we can get all the decision variable values as a component of our result object, `result0`. When `ompr` solves a problem, it creates an object with a number of components. After solving a model successfully in the console and saving the result to an object, you can see all the components by typing the object name and a dollar sign $. For example, using the command `Run All Code Chunks Above` and then typing `result0$` will show the following options.

- `model`: Saves the model that was solved into the result object.
- `objective_value`: Gives the optimal objective function value found.
- `status`: States whether the problem is optimal or not optimal.
- `solution`: An alphabetical listing of variables and their values in the optimal solution.

- `solution_column_duals`: Used in sensitivity analysis (Chapter 4)
- `solution_row_duals`: Also used in sensitivity analysis (Chapter 4)

Using this way of examining the solution (`result0$solution`) is particularly straightforward in an explicit model like this and where variables are conveniently alphabetized. Another approach is extract each variable's solution separately using the `get_solution` function from `ompr`.

```
kbl(res0, booktabs=T,
    caption="Base Case Two Variable Solution") |>
  kableExtra::kable_styling(latex_options = "hold_position")
```

TABLE 2.2 Base Case Two Variable Solution

	Profit	Ants	Bats
Solution	1980	180	60

It might be helpful to see what the model looks like before examining the results.

```
model0
```

```
## Mixed integer linear optimization problem
## Variables:
##   Continuous: 2
##   Integer: 0
##   Binary: 0
## Model sense: maximize
## Constraints: 4
```

Furthermore, we can use the `extract_constraints` command to see what the actual constraints look like in the model.

```
extract_constraints(model0)
```

```
## $matrix
## 4 x 2 sparse Matrix of class "dgCMatrix"
##
## [1,] 1  4
## [2,] 3  6
## [3,] 2  2
## [4,] 2 10
##
## $sense
## [1] "<=" "<=" "<=" "<="
##
## $rhs
## [1]  800  900  480 1200
```

This should look familiar. Let's see if we can arrange to make it look more similar to what we had given the model. Note that the term `RHS` represents the right-hand side of the constraint.

```
constr0<-extract_constraints(model0)
# Extract the constraints from model
constr00<- cbind(as.matrix(constr0$matrix), # Get matrix of data
                 as.matrix(constr0$sense),  # Get inequalities
                 as.matrix(constr0$rhs))    # Get right hand sides
rownames(constr00) <-c ('Machining','Assembly','Testing','Sensors')
kableExtra::kbl(constr00, booktabs=T, align='cccc',
```

```
  caption="Elements of the Constraints",
  col.names = c('Ants', 'Bats','Relationship','RHS'))|>
kableExtra::kable_styling(latex_options = "hold_position")
```

TABLE 2.3 Elements of the Constraints

	Ants	Bats	Relationship	RHS
Machining	1	4	<=	800
Assembly	3	6	<=	900
Testing	2	2	<=	480
Sensors	2	10	<=	1200

In the first version of the LP model implementation, we created many different versions of the model along the way while we kept adding elements to create a new model. This uses extra memory and may be a little slower. The second implementation keeps reusing the same model object until the very end when solving is done, and it places the results into an object. This may be more memory efficient and carries along with some extra notation.

2.4.2 Implementing the Base Case with Piping

In each step, we are simply taking the output of the previous step and feeding it in as the first term of the following step. It is almost like the output is a bucket that is then passed as a bucket into the next step. Rather than carrying a bucket from one step into the next one, a plumber might suggest a pipe connecting one step to the next. This is done so often that we refer to it as piping and a pipe operator. The **pipe symbol**, represented by `|>` will let us do this more compactly and efficiently. It takes a little getting used to but is a better way to build the model. Also, note that the piping operator requires `R` version 4.1.0. or higher.

Here is the equivalent process for building the base case implementation using a piping operator without all the intermediate partial models being built. Notice how it is a lot more concise. We will typically use this approach for model building, but both approaches are functionally equivalent.

```
result0 <- MIPModel()                                    |>
  add_variable(Ants, type = "continuous", lb = 0) |>
  add_variable(Bats, type = "continuous", lb = 0) |>
```

```
set_objective(7*Ants + 12*Bats, "max")            |>

add_constraint(1*Ants + 4*Bats<= 800)             |> # machining
add_constraint(3*Ants + 6*Bats<= 900)             |> # assembly
add_constraint(2*Ants + 2*Bats<= 480)             |> # testing
add_constraint(2*Ants + 10*Bats<= 1200)           |> # sensors
solve_model(with_ROI(solver = "glpk"))
```

Piping can make your code more readable by:

- structuring sequences of data operations left-to-right (as opposed to from the inside and out),
- avoiding nested function calls,
- minimizing the need for local variables and function definitions, and
- making it easy to add steps anywhere in the sequence of operations.

Piping's major drawback is that it treats all of the piped code as one line of code, and it can, therefore, make debugging much harder. The first line creates the basic model. The `|>` serves as a pipe symbol at the end of each line. It basically means take product of the previous command and feed it in as the first argument of the following command. While it may shorten each line of code and may be faster, there is a drawback in that the piped commands are treated as one very long line of code making it harder to find where an error occurs.

Let's check to see the status of the solver. Did it find the optimal solution? We do this by extracting the solver status from the result object.

```
print(solver_status(result0))
```

```
## [1] "optimal"
```

Furthermore, we can do the same thing to extract the objective function value.

```
print(objective_value(result0))
```

```
## [1] 1980
```

The command `objective_value(result0)` extracts the numerical result of the objective function (i.e. maximum possible profit), but it does not tell us what decisions give us this profit.

Since the LP solver found an optimal solution, let's now extract the solution values of the decision variables that give us this profit.

```
print(get_solution(result0, Ants))
```

```
## Ants
##  180
```

```
print(get_solution(result0, Bats))
```

```
## Bats
##   60
```

In summary, our optimal production plan is to make a mix of Ant and Bat drones. More specifically, we should produce 180 Ants and 60 Bats to generate a total profit of 1980. Given the situation, this is the optimal or most profitable possible production plan.

2.5 Adding a Third Product (Variable)

As always, it is important to start with formulating the model before moving into implementation. While this problem is small and straightforward, it is an important habit to develop as we build towards more complex models in later chapters.

2.5.1 Three Variable Base Case Formulation

We will now extend the previous model to account for a third product, the Cat drone. The goal is still to find the most profitable production plan. See the table to the right with a summary of the new situation.

TABLE 2.4 Three Variable Base Case

	Ant	Bat	Cat	Available
Profit	$7	$12	$5	
Machining	1	4	2	800
Assembly	3	6	2	900
Testing	2	2	1	480
Sensors	2	10	2	1200

A simple LP now is to find the production plan or amount of each of the products that results in the most profit. This will require a new decision variable, Cats, to be added to the model.

Let's extend our previous model. Just to reinforce the point, it is an important habit to very clearly define the decision variables.

- Ants = # of Ants to Make
- Bats = # of Bats to Make
- Cats = # of Cats to Make

Our objective function and constraints can now be written as an optimization model.

$$
\begin{aligned}
\text{Max } & 7 \cdot Ants + 12 \cdot Bats + 5 \cdot Cats \\
\text{s.t.: } & 1 \cdot Ants + 4 \cdot Bats + 2 \cdot Cats \leq 800 \\
& 3 \cdot Ants + 6 \cdot Bats + 2 \cdot Cats \leq 900 \\
& 2 \cdot Ants + 2 \cdot Bats + 1 \cdot Cats \leq 480 \\
& 2 \cdot Ants + 10 \cdot Bats + 2 \cdot Cats \leq 1200 \\
& Ants,\ Bats,\ Cats \geq 0
\end{aligned}
$$

2.5.2 Three Variable Base Case Implementation

Let's now implement the three variable case.

We have already loaded the required packages, so it is not necessary to reload them, and we can proceed directly into setting up the model.

```
model1 <- MIPModel()                                  |>
  add_variable(Ants, type = "continuous", lb = 0) |>
  add_variable(Bats, type = "continuous", lb = 0) |>
  add_variable(Cats, type = "continuous", lb = 0) |>
```

```
  set_objective(7*Ants + 12*Bats + 5*Cats,"max")  |>

  add_constraint(1*Ants + 4*Bats + 2*Cats<=800)   |>  # machining
  add_constraint(3*Ants + 6*Bats + 2*Cats<=900)   |>  # assembly
  add_constraint(2*Ants + 2*Bats + 1*Cats<=480)   |>  # testing
  add_constraint(2*Ants + 10*Bats + 2*Cats<=1200)     # sensors

result1 <-  solve_model(model1, with_ROI(solver="glpk"))
```

2.5.3 Three Variable Case Results and Interpretation

Let's check to see the status of the solver and the results.

```
print(solver_status(result1))
```

```
## [1] "optimal"
```

```
result1$objective_value
```

```
## [1] 2225
```

```
result1$solution
```

```
## Ants Bats Cats
##   50    0  375
```

Adding a decision variable, in this case, *Cats* enables new possibilities so it could increase profit.

2.6 Linear Programming Special Cases

There are several special cases where a linear program does not give the simple, unique solution that we might expect. These are:

- No feasible solution
- Multiple optima
- Redundant constraint
- Unbounded solution

Now, let's look at how we would modify the earlier formulation to come up with each of these situations.

2.6.1 Case 1: No Feasible Solution

Let's assume that the sales manager comes in and says that we have a contractual requirement to deliver 400 Ants to customers. This results in the following LP. [3]

$$
\begin{aligned}
\text{Max } & 7 \cdot Ants + 12 \cdot Bats + 5 \cdot Cats \\
\text{s.t.: } & 1 \cdot Ants + 4 \cdot Bats + 2 \cdot Cats \leq 800 \\
& 3 \cdot Ants + 6 \cdot Bats + 2 \cdot Cats \leq 900 \\
& 2 \cdot Ants + 2 \cdot Bats + 4 \cdot Cats \leq 480 \\
& 2 \cdot Ants + 10 \cdot Bats + 1 \cdot Cats \leq 1200 \\
& Ants \geq 400 \\
& Ants,\ Bats,\ Cats \geq 0
\end{aligned}
$$

Now let's extend our formulation with this change. In this case, we are going to simply add the new constraint to `model1` and create a new model, `model1infeas` to solve.

```
model1infeas <-
  add_constraint(model1, Ants >= 400) #THIS IS THE NEW CHANGE

result1infeas <- solve_model(model1infeas,
                             with_ROI(solver = "glpk"))
```

Note that the constraint on the minimum number of Ants could also be implemented by changing the lower bound on the `Ants` variable to be 400 instead of zero.

[3] It can be helpful to highlight a part of a LaTeX formulation. In this case, we can use `\textcolor{red}{` before the part that we want to turn red and use a closing `}`. You can also review the RMD file to see how it is done.

```
print(solver_status(result1infeas))
```

```
## [1] "infeasible"
```

```
result1infeas$solution
```

```
## Ants Bats Cats
##  240    0    0
```

Notice that since the solver status was infeasible, the values for the decision variables are not feasible and, therefore, cannot be considered a reliable (or possible) production plan. This highlights why the solver's status should always be confirmed to be "Optimal" before results are discussed.

2.6.2 Case 2: Multiple Optima

When a linear program is solved to optimality, it is an assurance that there is no better solution that can be found in terms of an objective function value. It is not a guarantee of being the only way of finding that good of a solution, though. The case of different decision variable values resulting in the same optimal objective function value is referred to as multiple optimal solutions.

There are a couple of ways of creating situations for multiple optima. One situation is to have a decision variable be identical or a linear multiple of another variable. In this case, each Cat now consumes exactly double the resources as an Ant and generates double the profit of an Ant. The new LP is shown in the following formulation.

$$
\begin{split}
\text{Max } & 7 \cdot Ants + 12 \cdot Bats + 14 \cdot Cats \\
\text{s.t.: } & 1 \cdot Ants + 4 \cdot Bats + 2 \cdot Cats \leq 800 \\
& 3 \cdot Ants + 6 \cdot Bats + 6 \cdot Cats \leq 900 \\
& 2 \cdot Ants + 2 \cdot Bats + 4 \cdot Cats \leq 480 \\
& 2 \cdot Ants + 10 \cdot Bats + 4 \cdot Cats \leq 1200 \\
& Ants, \; Bats, \; Cats \geq 0
\end{split}
$$

The implementation can be simplified again since we are only changing the objective function; let's change the objective function in `model1` and save it to `model2a`.

```
model2a <- MIPModel()                                   |>
  add_variable(Ants, type = "continuous", lb = 0) |>
  add_variable(Bats, type = "continuous", lb = 0) |>
  add_variable(Cats, type = "continuous", lb = 0) |>

  set_objective(7*Ants + 12*Bats + 14*Cats,"max") |>

  add_constraint(1*Ants + 4*Bats + 2*Cats<=800)   |> # machining
  add_constraint(3*Ants + 6*Bats + 6*Cats<=900)   |> # assembly
  add_constraint(2*Ants + 2*Bats + 4*Cats<=480)   |> # testing
  add_constraint(2*Ants + 10*Bats + 4*Cats<=1200)    # sensors

result2a <-  solve_model(model2a, with_ROI(solver="glpk"))
```

```
print(solver_status(result2a))
```

```
## [1] "optimal"
```

```
res2a <- cbind(objective_value (result2a),
               get_solution(result2a, Ants),
               get_solution(result2a, Bats),
               get_solution(result2a, Cats))
colnames(res2a)<-list("Profit","Ants","Bats","Cats")
rownames(res2a)<-list("Solution 2a")
```

```
kbl(res2a, booktabs=T,
    caption="First Solution of Multiple Optima Case") |>
  kableExtra::kable_styling(latex_options = "hold_position")
```

TABLE 2.5 First Solution of Multiple Optima Case

	Profit	Ants	Bats	Cats
Solution 2a	1980	180	60	0

Okay. When I ran it, all the production was focused on a mix of Ants and Bats but as discussed earlier, I think that there is an alternate solution producing Cats with the same total profit. The LP engine won't necessarily tell you that there is an alternate optimal solution. Let's see if we can "trick" the LP to show an alternate solution by preventing the production of any Ants by adding a constraint, $Ants = 0$ to `model2a` and naming this `model2b` and the solved object or solution as `result2b`. Notice that in a constraint, an equality constraint uses `==` rather than just `=`.

```
model2b <- add_constraint(model2a, Ants == 0)
          # FORCING LP TO FIND A DIFFERENT SOLUTION
result2b <- solve_model(model2b, with_ROI(solver = "glpk"))
print(solver_status(result2b))
```

```
## [1] "optimal"
```

```
res2b <- cbind(objective_value(result2b),
               get_solution(result2b, Ants),
               get_solution(result2b, Bats),
               get_solution(result2b, Cats))
colnames(res2b)<-list("Profit","Ants","Bats","Cats")
rownames(res2b)<-list("Solution 2b")
kbl (res2b, booktabs=T,
     caption="An Alternate Optimal Solution") |>
  kableExtra::kable_styling(latex_options = "hold_position")
```

TABLE 2.6 An Alternate Optimal Solution

	Profit	Ants	Bats	Cats
Solution 2b	1980	0	60	90

Again, a product mix is made but now instead and Ants and Bats, the mix is made up of Bats and Cats with exactly the same level of profit. This is an instance of multiple optima. Let's try one more situation by adding a constraint forcing the number of Ants to be 60 and solving.

```
model2c <- add_constraint(model2a, Ants == 60)
          # FORCING LP TO FIND A DIFFERENT SOLUTION
result2c <- solve_model(model2c, with_ROI(solver = "glpk"))
print(solver_status(result2c))
```

```
## [1] "optimal"
```

```
res2c <- cbind(objective_value(result2c),
               get_solution(result2c, Ants),
               get_solution(result2c, Bats),
               get_solution(result2c, Cats))
```

Let's summarize these results more clearly by displaying them in a single table.

TABLE 2.7 Examples of Alternate Optimal Solutions

	Profit	Ants	Bats	Cats
Solution 2a	1980	180	60	0
Solution 2b	1980	0	60	90
Solution 2c	1980	60	60	60

Note that all solutions generate the same profit.

Each of three solutions has exactly the same profit but different production plans to generate this profit. Furthermore, while we are listing three, there are actually many more solutions. You can force the system to find other solutions by setting the number of *Ants* to a number between 0 and 180 or *Cats* between 0 and 90. The number of solutions is infinite if we explore fractional solutions. More generally, when there are two alternate optimal in a linear program with continuous variables, there is actually an infinite number of other optimal solutions between them.

While our first optimization found a solution focused more on Ants, and our second found a solution that emphasized *Cats*, this is dependent upon the particular solver's algorithmic implementation. The user can consider this to be relatively arbitrary as there is no difference between them based on the objective function.

Which solution is the best? Within the limits of this problem, we can't distinguish between them and they are equally good. If an application area expert or the end decision maker prefers one solution over the other-refinements in costs or profits could be included directly, or additional constraints could be added. Sometimes considering these two otherwise equal solutions will elicit a response such as, "With all other things being equally, we prefer this alternative because of future market positioning." This could represent an additional or secondary goal, which is addressed using goal programming as discussed in Chapter 8.

2.6.3 Case 3: Redundant Constraint

For the redundant constraint, a new constraint for painting is created. Let's assume each item is painted and requires one liter of paint. We have 500 liters. The corresponding constraint is then added to the model.

$$
\begin{aligned}
\text{Max } & 7 \cdot Ants + 12 \cdot Bats + 5 \cdot Cats \\
\text{s.t.: } & 1 \cdot Ants + 4 \cdot Bats + 2 \cdot Cats \leq 800 \\
& 3 \cdot Ants + 6 \cdot Bats + 2 \cdot Cats \leq 900 \\
& 2 \cdot Ants + 2 \cdot Bats + 1 \cdot Cats \leq 480 \\
& 2 \cdot Ants + 10 \cdot Bats + 2 \cdot Cats \leq 1200 \\
& Ants + Bats + Cats \leq 500 \\
& Ants,\ Bats,\ Cats \geq 0
\end{aligned}
$$

Now we can implement the model. Rather than building the model from scratch, let's just add this one constraint to a previously built model.

```
model1redund <- add_constraint(model1, Ants + Bats + Cats <= 500)
        # THIS IS THE NEW CHANGE
```

```
result3 <- solve_model(model1redund, with_ROI(solver = "glpk"))
print(solver_status(result3))
```

```
## [1] "optimal"
```

```
result3$solution
```

```
## Ants Bats Cats
##   50    0  375
```

This constraint was *redundant* because the other constraints would keep us from ever having 500 drones or, therefore, ever needing 500 liters of paint. In other words, there is no way that this constraint could ever be binding at any solution regardless of what the objective function is. More precisely, the elimination of a redundant constraint does not change the size of the feasible region at all.

Note that not all non-binding constraints at an optimal solution are redundant. Deleting a non-binding constraint and resolving won't change the optimal objective function value. On the other hand, for a different objective function, that non-binding constraint might become binding, and therefore, different solutions would be found if it were deleted.

Challenge 1: Can you use a calculator to simply estimate the maximum number of Ants that could be made? Bats? Cats?
Challenge 2: How would you modify the formulation to find the most total drones that could be produced irrespective of profit?

2.6.4 Case 4: Unbounded Solution

As with other cases, there are multiple ways of triggering this condition. For the unbounded solution, instead of at *most* a certain amount of resources can be used, the constraints are changed to at *least* that amount of each resource must be used. This doesn't make a lot of sense in the setting of this application. Perhaps a cynic would say that in a cost-plus business arrangement or a situation where the factory manager has a limited purview and doesn't see issues such as downstream demand limits and cost impacts, it results in this kind of myopic perspective. More commonly, an unbounded solution might be a sign that the analyst had simply reversed one or more inequalities or the form of the objective (max vs. min).

$$
\begin{split}
\text{Max } 7 \cdot Ants + 12 \cdot Bats + 5 \cdot Cats \\
\text{s.t.:} \\
1 \cdot Ants + 4 \cdot Bats + 2 \cdot Cats \geq 900 \\
3 \cdot Ants + 6 \cdot Bats + 2 \cdot Cats \geq 800 \\
2 \cdot Ants + 2 \cdot Bats + 1 \cdot Cats \geq 480 \\
2 \cdot Ants + 10 \cdot Bats + 2 \cdot Cats \geq 1200 \\
Ants, Bats, Cats \geq 0
\end{split}
$$

Let's make this change to the model and the implementation by simply changing each $\leq$ to a $\geq$ for each constraint.

```
result4 <- MIPModel()                                    |>
 add_variable(Ants, type = "continuous", lb = 0) |>
 add_variable(Bats, type = "continuous", lb = 0) |>
 add_variable(Cats, type = "continuous", lb = 0) |>

 set_objective(7*Ants + 12*Bats + 5*Cats,"max")   |>
```

```
add_constraint(1*Ants + 4*Bats + 2*Cats>= 800)  |> # machining
add_constraint(3*Ants + 6*Bats + 2*Cats>= 900)  |> # assembly
add_constraint(2*Ants + 2*Bats + 1*Cats>= 480)  |> # testing
add_constraint(2*Ants + 10*Bats + 2*Cats>= 1200)|> # sensors
solve_model(with_ROI(solver = "glpk"))
```

Now let's see what is reported from trying to solve this model.

```
print(solver_status(result4))
```

```
## [1] "infeasible"
```

```
result4$solution
```

```
## Ants Bats Cats
##    0  240    0
```

The solver status reports that the problem is *infeasible* rather than *unbounded*, but by inspection, the solution is feasible in that it satisfies all of the constraints of $\geq$ and therefore, the LP is feasible.

Infeasible vs. Unbounded. This is a known issue in `ompr` as of 0.8.1 and reported on github. It is caused by not being able to distinguish between the different status conditions for situations other than being solved to guaranteed optimality. This is caused by the variety of status codes used by different LP solvers (`glpk`, `symphony`, and others) passing a numerical status code to `ROI`. Each solver has defined their status codes in different ways.) The result is that you should read the `ompr` status of "infeasible" to indicate that is not assured of being an optimal solution. The misinterpretation of solver status is an outcome of the independent toolchain from the solver (`glpk` to `ROI` to `ompr`.)

This is another good reminder that it is important to always check the status of the solver.[4]

---

[4]ompr 1.0 implemented a fix to change the solver status to "error" to indicate that it did not solve the problem to optimality. Detailed information is available from result4$additional_solver_output

2.7 Abstracting the Production Planning Model

We have explicitly created a two-variable model and a three-variable model by naming each variable independently. This process doesn't scale well for companies with dozens, hundreds, or thousands of different products. Simply writing out the full linear program gets very tedious, hard to read, and even maintain. An application for a company with a thousand products and a thousand resources would have a million terms. Assume that variables are on average seven letters long, each resource consumed is a single-digit whole number, and a plus symbol is used to add terms together and no spaces. This means that there will be $(7+1) \cdot 1000 + 999 = 8999$ characters in each line before the inequality. Just say each constraint has 9000 characters. If a line has 60 characters, this would mean 150 lines or around two pages for each resource (constraint.) The 1000 resources would correspond to about 2000 pages, along with an objective function and non-negativity constraints. All in all, this single model would make for some rather dry reading.

In practice, people don't write out the full LP explicitly for large models. This includes journals, no journal's page limit would be able to accommodate the above explicit linear program even if readers had the patience to wade through the model.

Rather than writing out models explicitly, instead we should express them algebraically. The products are numbered instead of given names: Ants, Bats, and Cats become products 1, 2, and 3, respectively.

We could adopt a variety of naming or notation conventions of the variables:

- Ants, Bats, Cats
- Product1, Product2, Product3
- X1, X2, X3
- `X[1], X[2], X[3]`
- X_1, X_2, X_3
- $X_i,\ i = 1, ..., 3$

Each of these conventions can have its place.

The notation with brackets, `X[1]` is consistent R notation and that of many other computer languages. This allows us to simply use a vector of X where we can use each element of the vector for each of the products to use. This connects very well with the data structures available to us in R (and other languages.) It would also allow us to handle any number of products. If we had a thousand products, the thousandth product is simply `X[1000]`.

Subscripting makes for a more succinct and compact way of expressing the same concept.

Similarly, the resources: Machining, Assembly, Testing, and Sensors resources are numbered as 1, 2, 3, and 4, respectively. Note that we do not need to separate the resources by units; the first three are in units of hours, while the last is a count of sensors.

Here we are talking about abstracting the model in terms of variable names and notation. In the next chapter, we will continue with generalizing the model's number of products and resources.

2.8 Methods of Solving Linear Programs

In this book, we will focus on solving linear programs using the Simplex method. While we don't cover the algorithm of the Simplex method, it is well implemented in a variety of robust packages that we can readily use for modeling purposes. Conceptually, the Simplex method can be thought of as traversing the edges of a multi-dimensional space. In general, it is usually a very quick and efficient process for even very large problems. There are alternatives to solving optimization problems.

First, Karmarkar developed an approach called Interior Points Algorithm that instead of following the edges, instead cuts through the middle of the multi-dimensional region. On certain problems, this can be significantly faster. Often a standard linear programming solver will include an option to use an interior points approach.

Another very important way of solving optimization problems is using heuristic methods. These approaches cover a range of techniques but in general, unlike the Simplex and Interior Points method approaches do not guarantee optimality. These include evolutionary or genetic algorithms, simulated annealing, and gradient search (hill-climbing). These approaches can be particularly helpful on nonlinear problems or complex integer programming problems.

Changing the method of solving an optimization problem can often be done at the implementation stage after the model is formulated. The approach of formulating the model will often stay the same regardless of the approach used for solving.

2.9 Exercises

Exercise 2.1 (Adding a Dog Drone). Your company has extended production to allow for producing the dog drone and is now including a finishing department that primes and paints the drones.

a. Use R Markdown to create your own description of the model.
b. Extend the R Markdown to show your LP Model. Be sure to define models.
c. Solve the model in R.
d. Interpret and discuss the model in R Markdown.
e. Discuss how one parameter would need to change in order to result in a different production plan. Demonstrate how this affects the results.

TABLE 2.8 Data for Dog Drone Exercise

Characteristic	Ants	Bats	Cats	Dogs	Available
Profit	\$7	\$12	\$5	\$24	
Machining	1	4	2	2	800
Assembly	3	6	2	2	900
Testing	2	2	1	2	480
Sensors	2	10	2	4	1200
Painting	2	7	2	6	500

Hint: Knit your RMarkdown as a PDF. Also, don't be bothered by fractional values for a possible production plan at this time. It is only a high-level production plan so if a drone is only partially finished, assume that work on them will continue in the next planning period. We will deal with eliminating fractional variable values much more formally in Chapter 6, but for now, consider the expression "close enough for government work" to apply.

Exercise 2.2 (Eliminating Painting of Bat and Dog Drones). Exercise 2.1b (Eliminating Painting Cost) Your company has changed production plans for producing the cat and dog drone models so that they no longer require painting. However, a market analysis predicts a decrease in profitability to \$18 for each dog drone,

TABLE 2.9 Characteristics with Painting Changes

Characteristic	Ants	Bats	Cats	Dogs	Available
Profit	$7	$10	$5	$18	
Fabrication	1	3	2	2	800
Assembly	3	4	2	2	900
Machining	2	3	1	2	480
Sensors	3	5	2	4	1200
Painting	2	0	2	0	500

a. Use R Markdown to create your own description of the model.
b. Extend the R Markdown to show your LP Model.
c. Solve the model in R.
d. Interpret and discuss the model in R Markdown.
e. Discuss about objective value, how it has changed from the previous results. How this will help you in future to calculate things and decisions.

Exercise 2.3 (Printing Solutions). From the above exercise solution, fetch results of each drone for comparing each drone's production plan before and after the change so that the reflection of changes in production plan can be discussed. This will be handy in using library functions explained in the chapter and will help in obtaining required results in later exercises in this book.

Exercise 2.4 (Rose City Roasters). Rose City Roasters provides premium roasted coffee beans for regional coffee shops. There are four important steps: Arranging the beans, roasting the beans, allowing them to cool, and then giving them a chance to cool down. This is followed by degassing and then packaging.

TABLE 2.10 Data for Rose City Roasters

<table>
<tr><th></th><th>Light</th><th>Medium</th><th>Medium-Dark</th><th>Dark</th><th>Available (hours)</th></tr>
<tr><td>Price</td><td>$5.0</td><td>$5.2</td><td>$5.55</td><td>$6</td><td></td></tr>
<tr><td colspan="6">(/*lb*)|$8|$10|$9|$11|||*VariableCost*(/lb)</td></tr>
<tr><td>Arrange (lbs/hr)</td><td>10</td><td>10</td><td>10</td><td>10</td><td>4000</td></tr>
<tr><td>Roast (lbs/hr)</td><td>6</td><td>4</td><td>2</td><td>8</td><td>2000</td></tr>
<tr><td>Cool (lbs/hr)</td><td>3</td><td>2</td><td>2</td><td>2</td><td>1480</td></tr>
<tr><td>Degas (lbs/hr)</td><td>3</td><td>4</td><td>2</td><td>4</td><td>1200</td></tr>
<tr><td>Packaging (lbs/hr)</td><td>4</td><td>4</td><td>2</td><td>3</td><td>1200</td></tr>
</table>

Create an appropriate formulation to maximize profit (selling price minus variable cost).

Implement and solve the resulting model.

Discuss results.

3

More Linear Programming Models

3.1 Types of LP Models

In this chapter, we will examine a range of classic applications of linear programs. These applications will give ideas for how to model a variety of situations. In each case, try to follow along with the application.

3.2 The Algebraic Model

In the previous chapter, we examined situations with only a few products and constraints. In general, most companies have many more products, and we won't be wanting to name each variable explicitly and uniquely. We refer to this type of model as an *explicit linear program*. Instead, we use sets of products and resources.

We could also reframe the model more algebraically. Let's use subscripts to differentiate between products and resources. We can define to *i=1* to represent Ants, *i=2* to represent Bats, and *i=3* to represent Cats. Similarly, *j=1* represents machining, *j=2*, represents assembly, etc.

Now, let's move on to defining the data. Let's define the amount to produce of each product, *i*, as x_i and resource *j* consumed by product *i* as $R_{i,j}$. The available resource *j* is then A_j. The profit per product *i* is then P_i. The LP can now be rewritten as shown below with the three products and four constraints.

$$\begin{aligned} \text{Max } & \sum_{i=1}^{3} P_i x_i \\ \text{s.t.: } & \sum_{i=1}^{3} R_{i,j} x_i \leq A_j, \; j = 1, 2, 3, 4 \\ & x_1, \; x_2, \; x_3 \geq 0 \end{aligned}$$

DOI: 10.1201/9781003051251-3

We could further generalize this by eliminating the hard coding of the number of products and resource constraints. Instead we define the number of products and resources as *NProd* and *NResources*, respectively.

Let's introduce a very convenient shorthand symbol, $\forall$, is read as "for all." It can be interpreted to mean "repeat by substituting **for all** possible values of this subscript." In the constraint, it means to repeat the constraint line with $j = 1$, again with $j = 2$, and so on for as many values of j as make sense in the model.

We also will use it in the non-negativity constraint to avoid having to list out every variable, *x_1*, *x_2*, etc. In other words, given that i is used consistently for the three products, then $x_i \geq 0 \; \forall \; i$ is equivalent to $x_i \geq 0 \; , i = 1, 2, 3$ or even $x_1 \geq 0, \; x_2 \geq 0, \; x_3 \geq 0$.

The payoff is modest when there are 3 products and 4 constraints but when there are hundreds of products and thousands of constraint, the benefit is tremendous. Furthermore, the generalized, algebraic model doesn't change when a new product is added or one is discontinued, it is only a change of data that is fed into the model. The result is that the $\forall$ symbol can simplify the description of complex models when index ranges are clear.

We can then rewrite the previous formulation as the following, more general formulation.

$$
\begin{aligned}
\text{Max:} & \sum_{i=1}^{NProd} P_i x_i \\
\text{subject to} & \sum_{i=1}^{NProd} R_{i,j} x_i \leq A_j \; \forall \; j \\
& x_i \geq 0 \; \forall \; i
\end{aligned}
$$

It is even more important to clearly and precisely define the meanings of variable and data element in an algebraic model than in an explicit model.

3.2.1 Tips and Conventions for Algebraic Models

Learning how to read and write mathematical models is an important skill. In this section we will provide a quick overview. More information is provided in *Appendix B*.

Another good practice is to use a mnemonic to help suggest the meaning of data. That is why I chose "R" for Resource, "P" for Profit, and "A" for Available resources.

A helpful convention is to use capital letters for data and lower case letters for variables. Some people will swap this around and use capital letters for variables and lower case letters for data – it doesn't matter as long as a model is consistent. It gives a quick visual cue for the reader as to whether each item is a variable or constraint.

More complex models often run out of letters that make sense. A common approach in these models is to use a superscript.
For example, perhaps labor cost for each worker, w, could have different values for both normal and overtime rates. Rather than separate data terms for such closely related concepts, we might denote regular hourly labor cost for worker w as C_w^R and for overtime as C_w^O. Another option is to use a Greek letter such as θ. We'll see a few Greek letters used in chapter 5.

Again, this highlights why it is very important to clearly define all the data and variables used in the model. Think of it as building a language for describing the particular model. It is frequently an iterative process that it is updated as the model is refined and developed. Frequently a definitions section will warrant its own section or subsection and should precede the formulation.

3.2.2 Building the Generalized Model in R

3.2.2.1 Preparing the Data

The concise, algebraic representation can be easily scaled to any number of products and resources. I'll expand the names of data slightly for making the R code more readable but this is meant to be consistent with the above formulation.

TABLE 3.1 Profit per Product

	Prod1	Prod2	Prod3	Prod4
Profit	7	10	5	24

Let's implement a four product model requiring varying levels of five limited resources. We will start our implementation by creating the appropriate data structures in R.

```
NProd <- 4
NResources <- 5
ProdNames <- lapply(list(rep("Prod",NProd)),paste0,1:NProd)
              # Product names: Prod1, Prod2, ...
Profit <- matrix(c(20, 14, 3, 16),
                 ncol=NProd,dimnames=c("Profit",ProdNames))
```

```
ResNames<- lapply(list(rep("Res",NResources)),
                  paste0,1:NResources)
                  # Resource names: Res1, Res2, ...
Resources <- matrix(c( 1, 3, 2, 3, 2, 3, 4, 3, 5, 7,
                       2, 2, 1, 2, 2, 2, 2, 2, 4, 6),
            ncol=NProd,
            dimnames=c(ResNames,ProdNames))
Available <- matrix(c(800, 900, 480, 1200, 500),
            ncol=1,dimnames=c(ResNames,"Available"))
```

This should match the data that we hard coded into the R linear programming model in the previous chapter.

Similarly, we can display the resources used by each product and the amount of each resource available alongside each other.

```
Combined <- cbind(Resources, Available)
kbl(Combined, booktabs=T,
    caption="Resources Used by Each Product and Amount Available") |>
  kable_styling(latex_options = "hold_position")
```

TABLE 3.2 Resources Used by Each Product and Amount Available

	Prod1	Prod2	Prod3	Prod4	Available
Res1	1	3	2	2	800
Res2	3	4	2	2	900
Res3	2	3	1	2	480
Res4	3	5	2	4	1200
Res5	2	7	2	6	500

We have used the `cbind` function to do a column binding of the data. In this way the representation is more visually intuitive.

To ensure that we know how to access the data, if we want to see how the amount of the first resource used by the second product, you can enter `Resources[1,2]` in R Studio's console, which is then evaluated as 3.

Using inline code is a great feature of RMarkdown. In the text of your RMarkdown, anything enclosed by a pair of single tick marks will be shown as a code chunk. If it starts with the letter `r`, it will be passed instead to R for evaluation. This allows for discussing results in a reproducible manner. For example, rather than discussing a result giving a value of 42.00 in the text and a later rerun of the analysis has an updated table of results showing the result to be 42.42, by using inline code evaluation, it will always show the up to date result.

Now, let's begin building our optimization model.

3.2.2.2 Implementing the Model

First, we'll start by loading the packages that we are using. I'll typically do this as at the beginning of an RMarkdown document but for demonstration purposes, I'll put it here at the beginning of our model implementation. It is important to note that when knitting an RMarkdown document, it will run as a fresh environment without having items in memory from your environment and packages loaded. When you run all code chunks, it will retain the currently loaded packages.

```
suppressPackageStartupMessages(library (dplyr, quietly = TRUE))
suppressPackageStartupMessages(library (ROI, quietly = TRUE))
library (ROI.plugin.glpk, quietly = TRUE)
library (ompr, quietly = TRUE)
library (ompr.roi, quietly = TRUE)
```

We will continue with the code to build the model in a generic format. Note that in my `ompr` model, I generally like to give each linear programming variable in `ompr` a `V` prefix to differentiate it from a regular `R` data object that might exist elsewhere in my `R` environment. Conflicts between `R` objects and `ompr` variables are a common problem for modelers and this helps to avoid the problem, assuming that you don't have a lot of other `R` objects that start with `V`. For more discussion about this issue and other common problems that are particularly common in optimization modeling using R, see Appendix C.

```
prodmodel <- MIPModel()                                 |>
  add_variable (Vx[i], i=1:NProd,
                type="continuous", lb=0)                |>
  set_objective (sum_expr(Profit[i] * Vx[i],
                     i=1:NProd ), "max")                |>
  add_constraint (sum_expr(Resources[j,i]*Vx[i],
                     i=1:NProd)
                 <= Available[j],
                 j=1:NResources)                        |>
```

```
  solve_model(with_ROI(solver = "glpk"))

prodmodel
```

```
## Status: optimal
## Objective value: 4800
```

3.2.3 Examining the Results

Displaying the object of `prodmodel` only shows a simple summary of the results of the analysis. It indicates whether the model was solved to optimality (and it was!) and the objective function value (profit).

This is useful to know but we are really interested in how to generate this profit. To do this, we need to extract the values of the variables.

```
results.products <- matrix (rep(-1.0,NProd),
                           nrow = NProd, ncol=1,
                           dimnames=c(ProdNames,c("x")))
temp <- get_solution (prodmodel, Vx[i])
  # Extracts optimal values of variables
results.products <- t(temp [,3] )
  #Extracts third column
results.products <- matrix (results.products,
                           nrow = 1, ncol=NProd,
                           dimnames=c(c("x"),ProdNames))
                           # Resizes and renames
kbl(format(head(results.products),digits=4), booktabs=T,
    caption="Optimal Production Plan") |>
  kable_styling(latex_options = "hold_position")
```

TABLE 3.3 Optimal Production Plan

	Prod1	Prod2	Prod3	Prod4
x	240	0	0	0

The table displays the optimal production plan.

Let's examine how the resources are consumed. To do this, we can multiply the amount of each product by the amount of each resource used for that product. For the first product, this would be a term by term sum of each product resulting in 240, which is less than 800. We can do this manually for each product. Another approach is to use the command, `Resources[1,]%*%t(results.products)`. This command will take the first row of the Resources matrix and multiplies it by the vector of results.

Note how we used both an inline r expression and an inline executable r code chunk. Inline code chunks can be inserted into text by using a single tick at the beginning and end of the chunk instead of the triple tick mark for regular code chunks. Also, the inline r code chunk starts with the letter r to indicate that it is an r command to evaluate. A common use for this might be to show the results of an earlier analysis.

One thing to note is that the first row of `Resources` is by definition a row vector and `result.products` is also a row vector. What we want to do is do a row vector multiplied by a column vector. In order to do this, we need to convert the row vector of results into a column vector. This is done by doing a *transpose* which changes the row to a column. This is done often enough that the actual function in `R` is just a single letter, `t`.

We can go one further step now and multiply the matrix of resources by the column vector of production.

```
results.Resources <- Resources[]%*%t(results.products)
# Multiply matrix of resources by amount of products produced
ResourceSlacks <- cbind (results.Resources,
                         Available,
                         Available-results.Resources)
colnames(ResourceSlacks)<-c("Used", "Available", "Slack")
kbl(format(head(ResourceSlacks),digits=4), booktabs=T,
    caption="Resources Used in Optimal Solution") |>
  kable_styling(latex_options = "hold position")
```

This section covered a lot of concepts including defining the data, setting names, using indices in `ompr`, building a generalized `ompr` model, extracting decision variable values, and calculating constraint right-hand sides. If you find this a little uncomfortable, try doing some experimenting with the model. It may take some experimenting to get familiar and comfortable with this.

TABLE 3.4 Resources Used in Optimal Solution

	Used	Available	Slack
Res1	240	800	560
Res2	720	900	180
Res3	480	480	0
Res4	720	1200	480
Res5	480	500	20

3.2.4 Changing the Model

Let's modify the above model. We can do this by simply changing the data that we pass into the model and rebuilding the model.

Let's change the number of sensors required for an Ant to 5. Recall that this is the first product and the fourth resource. This is how we can change the value.

```
Resources[4,1] <- 5
# Set value of the 4th row, 1st column to 5
# In our example, this is the number of sensors needed per Ant
```

Now we will rebuild the optimization model. Note that simply changing the data does not change the model.

```
prodmodel <- MIPModel() |>
  add_variable (Vx[i], i=1:NProd,
                type="continuous", lb=0)        |>
  set_objective (sum_expr(Profit[i] * Vx[i] ,
                          i=1:NProd ), "max")  |>
  add_constraint (sum_expr(Resources[j,i]*Vx[i],
                           i=1:NProd)
                  # Left hand side of constraint
                  <= Available[j],
                  # Inequality and Right side of constraint
                  j=1:NResources)              |>
                  # Repeat for each resource, j.
  solve_model(with_ROI(solver = "glpk"))

prodmodel
```

```
## Status: optimal
## Objective value: 4800
```

We can see that the objective function has changed.

```
results.products <- matrix (rep(-1.0,NProd),
                             nrow = NProd,
                             ncol=1,
                             dimnames=c(ProdNames,c("x")))
temp <- get_solution (prodmodel, Vx[i]) # Extracts optimal var. values
results.products <- t(temp [,3] )       # Extracts third column
results.products <- matrix (results.products,
                             nrow = 1,
                             ncol=NProd,
                             dimnames=c(c("x"),ProdNames))
   # Resizes and renames
kbl(results.products, booktabs=T,
   caption="Revised Optimal Production Plan")  |>
  kable_styling(latex_options = "hold_position")
```

TABLE 3.5 Revised Optimal Production Plan

	Prod1	Prod2	Prod3	Prod4
x	240	0	0	0

The production plan has significantly changed.

3.3 Common Linear Programming Applications

3.3.1 Blending Problems

Specific blend limitations arise in many situations. Recall from our three variable case in Chapter 2 that the vast majority of drones produced were Cat models. Let's assume that we can't have more than 40% of total production made up of Cats. Let's start by building our original three-variable model. In our original case, this would be expressed as the following.

$$
\begin{aligned}
\text{Max } & 7 \cdot Ants + 12 \cdot Bats + 5 \cdot Cats \\
\text{s.t.:} & \\
& 1 \cdot Ants + 4 \cdot Bats + 2 \cdot Cats \leq 800 \\
& 3 \cdot Ants + 6 \cdot Bats + 2 \cdot Cats \leq 900 \\
& 2 \cdot Ants + 2 \cdot Bats + 1 \cdot Cats \leq 480 \\
& 2 \cdot Ants + 10 \cdot Bats + 2 \cdot Cats \leq 1200 \\
& Ants, \ Bats, \ Cats \geq 0
\end{aligned}
$$

Let's rebuild our three variable implementation and review the results to see if it is violates our blending requirement.

```
model1 <- MIPModel()                                          |>
  add_variable(Ants, type = "continuous", lb = 0) |>
  add_variable(Bats, type = "continuous", lb = 0) |>
  add_variable(Cats, type = "continuous", lb = 0) |>

  set_objective(7*Ants + 12*Bats + 5*Cats,"max")  |>

  add_constraint(1*Ants + 4*Bats + 2*Cats<=800)   |> # machining
  add_constraint(3*Ants + 6*Bats + 2*Cats<=900)   |> # assembly
  add_constraint(2*Ants + 2*Bats + 1*Cats<=480)   |> # testing
  add_constraint(2*Ants + 10*Bats + 2*Cats<=1200)    # sensors

res3base <-  solve_model(model1, with_ROI(solver="glpk"))

xants <- get_solution (res3base, Ants)
xbats <- get_solution (res3base, Bats)
xcats <- get_solution (res3base, Cats)
base_case_res             <- cbind(xants,xbats,xcats)
rownames (base_case_res) <- "Amount"
colnames (base_case_res) <- c("Ants","Bats","Cats")
```

```
kbl(base_case_res, booktabs=T,
    caption="Production Plan for Base Case") |>
  kable_styling(latex_options = "hold_position")
```

These results clearly violate the required production mix for Cats. Let's now work on adding the constraint that Cats can't make up more than 40% of the total production.

TABLE 3.6 Production Plan for Base Case

	Ants	Bats	Cats
Amount	50	0	375

$$\frac{Cats}{Ants + Bats + Cats} \leq 0.4$$

Alas, this is not a linear function since we are dividing a variable by a function of variables so we need to clear the denominator. For demonstration purposes, we will write out each step.

$$Cats \leq 0.4 \cdot (Ants + Bats + Cats)$$

We like to get all the variables on the left side so let's move them over.

$$Cats - 0.4 \cdot Ants - 0.4 \cdot Bats - 0.4 \cdot Cats \leq 0$$

Let's simplify this a little, which gives us the following:

$$-0.4 \cdot Ants - 0.4 \cdot Bats + 0.6 \cdot Cats \leq 0$$

Now we can just this to the original formulation. The result is the following formulation.

$$\begin{aligned}
\text{Max } & 7 \cdot Ants + 12 \cdot Bats + 5 \cdot Cats \\
\text{s.t.:} & \\
& 1 \cdot Ants + 4 \cdot Bats + 2 \cdot Cats \leq 800 \\
& 3 \cdot Ants + 6 \cdot Bats + 2 \cdot Cats \leq 900 \\
& 2 \cdot Ants + 2 \cdot Bats + 1 \cdot Cats \leq 480 \\
& 2 \cdot Ants + 10 \cdot Bats + 2 \cdot Cats \leq 1200 \\
& -0.4 \cdot Ants - 0.4 \cdot Bats + 0.6 \cdot Cats \leq 0 \\
& Ants, \ Bats, \ Cats \geq 0
\end{aligned}$$

We can simply add a constraint to an existing `ompr` model of our 3 variable base case and then solve again.

```
ModelBlending<- add_constraint(model1,
              -0.4*Ants -0.4*Bats +0.6*Cats <= 0)
resBlending<- solve_model(ModelBlending,
                          with_ROI(solver = "glpk"))
```

```
blendres<-cbind(get_solution (resBlending, Ants),
                get_solution (resBlending, Bats),
                get_solution (resBlending, Cats))
```

Okay, now let's put both side by side in a table to show the results.

```
rownames(base_case_res)<-"Base Model"
rownames(blendres)<-"with Constraint"
comparative <- rbind(base_case_res,round(blendres, 2))
colnames (comparative) <- c("Ants","Bats","Cats")
kbl(comparative, booktabs=T,
    caption =
     "Compare Baseline and Production Plan with Blending Constraint")|>
  kable_styling(latex_options = "hold_position")
```

TABLE 3.7 Compare Baseline and Production Plan with Blending Constraint

	Ants	Bats	Cats
Base Model	50	0	375
with Constraint	140	40	120

The table clearly shows that the blending constraint had a major impact on our product mix.

3.4 Allocation Models

An allocation model divides resources and assigns them to competing activities. Typically it has a maximization objective with less than or equal to constraints. Note that our production planning problem from Chapter 2 is an allocation model.

$$\text{Max: } \sum_{i=1}^{3} P_i x_i \qquad \text{[Maximize Profit]}$$

$$\text{s.t. } \sum_{i=1}^{3} R_{i,j} x_i \leq A_j \; \forall \, j \quad \text{[Resource Limits]}$$

$$x_i \geq 0 \; \forall \, i$$

3.4.1 Covering Models

A covering model combines resources and coordinates activities. A classic covering application would be what mix of ingredients "covers" the requirements at the lowest possible cost. Typically it has a minimization objective function and greater than or equal to constraints.

$$\text{Min: } \sum_{i=1}^{3} C_i x_i \qquad \text{[Minimize Cost]}$$

$$\text{S.t. } \sum_{i=1}^{3} A_{i,j} x_i \geq R_j \; \forall \, j \quad \text{[Meet Requirements]}$$

$$x_i \geq 0 \; \forall \, i$$

Consider the case of Trevor's Trail Mix Company. Trevor creates a variety of custom trail mixes for health food fans. He can use a variety of ingredients displayed in following table.

TABLE 3.8 Trevor Trail Mix Company Ingredients

Characteristic	Mix1	Mix2	Mix3	Mix4	Min Req.
Cost	$20	$14	$3	$16	
Calcium	6	2	1	4	1440
Protein	8	6	1	8	1440
Carbohydrate	6	4	1	25	2000
Calories	7	10	2	12	1000

Let's go ahead and build a model in the same way as we had done earlier for production planning.

```
NMix <- 4
NCharacteristic <- 4
MixNames <- lapply(list(rep("Mix",NMix)),paste0,1:NMix)
```

```
                    # Mix names: Mix1, Mix2, ...
CharNames <-lapply(list(rep("Char",NCharacteristic)),
                    paste0,1:NCharacteristic)
                    # Characteristics of each mix
Cost <- matrix(c(20, 14, 3, 16),
               ncol=NMix,dimnames=c("Cost",MixNames))
MixChar <- matrix(c( 6, 8, 6, 7,
                     2, 6, 4, 10,
                     1, 1, 1, 2,
                     4, 8, 25, 12),
               ncol=4, dimnames=c(CharNames,MixNames))
CharMin <- matrix(c(1440, 1440, 2000, 1000),
               ncol=1,dimnames=c(CharNames,"Minimum"))
```

```
TTMix <-cbind(MixChar,CharMin)
kbl(TTMix, booktabs=T,
    caption="Data for Trevor Trail Mix Company") |>
  kable_styling(latex_options = "hold_position")
```

TABLE 3.9 Data for Trevor Trail Mix Company

	Mix1	Mix2	Mix3	Mix4	Minimum
Char1	6	2	1	4	1440
Char2	8	6	1	8	1440
Char3	6	4	1	25	2000
Char4	7	10	2	12	1000

Hint: You might need to add a total amount to make! Modify the numbers until it runs.

Now let's build our model.

```
trailmixmodel <- MIPModel()                                |>
 add_variable(Vx[i],i=1:NMix,type="continuous",lb=0)      |>
 set_objective(sum_expr(Cost[i]*Vx[i],i=1:NMix ),"min") |>
 add_constraint(sum_expr(MixChar[j,i]*Vx[i],i=1:NMix)  # LHS
                >= CharMin[j],              # Inequality and RHS
                j=1:NCharacteristic)        # Repeat for each resource
```

```
results.trailmix <- solve_model(trailmixmodel,
                                with_ROI(solver = "glpk"))

results.trailmix
```

```
## Status: optimal
## Objective value: 4426.667
```

```
xvalue <- t(get_solution(results.trailmix, Vx[i])[,3])
```

We'll leave it to the reader to clean up the output of results.

Another classic example of a covering problem is a staff scheduling problem. In this case, a manager is trying to assign workers to cover the required demands throughout the day, week, or month.

3.4.2 Transportation Models

A transportation model is typically for getting material from one place to another at the lowest possible costs. It has sets of source points or nodes as well as ending or destination nodes. The decision variables are the amount to send on each route. Constraints are typically based on supply from the source nodes and capacity at the destination nodes.

This naturally lends itself to potential network diagrams such as the one to the right. In this case we have three warehouses (or supply points) in Minnesota (MN), Pennsylvania (PA), and Wyoming (WY), each with a supply shown to the left. On the right side, we have four customer or demand nodes: Atlanta (ATL), Boston (Bos), Chicago (Chi), and Denver (Den) with their maximum demand shown to the right of the node. We also have arcs connecting every supply node to every demand node, each labeled with a corresponding cost per unit to transport an item along that arc. The goal is to get as much shipped as possible at the lowest possible cost.

Let's start to model this now. We want to define the decision variables, in this case, $x_{i,j}$ is the amount of product to ship from node i to node j.

Let's also define the data available. The cost per unit to ship from node i to node j is $C_{i,j}$. The supply available from each supply node is S_i and the maximum demand that can be accommodate from each destination node is D_j.

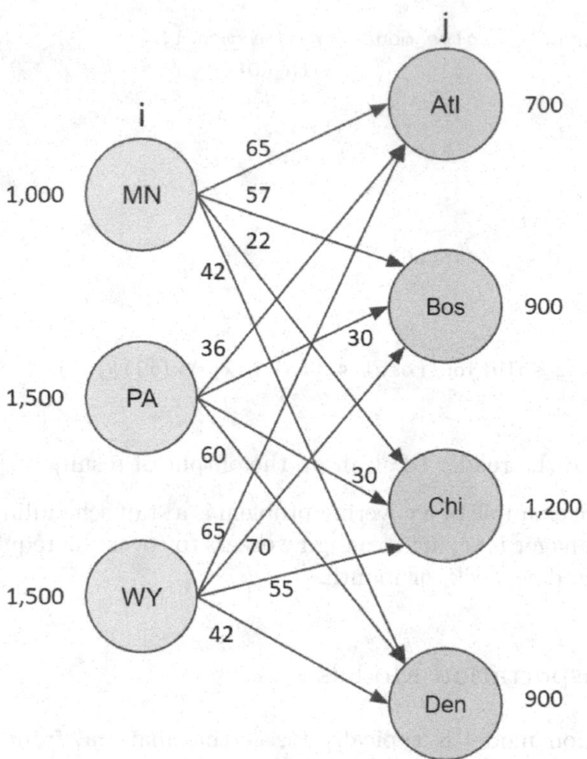

FIGURE 3.1 Transportation Example.

In this formulation we need to make sure that we don't ship out more than the capacity of each supply node.

Similarly, we need to ensure that we don't take in more than demand capacity at any destination.

$$
\begin{aligned}
\text{Min} & \sum_i \sum_j C_{i,j} x_{i,j} & \\
\text{s.t.} & \sum_i x_{i,j} \leq D_j \; \forall \, j & \text{[Demand Limits]} \\
& \sum_j x_{i,j} \leq S_i \; \forall \, i & \text{[Supply Limits]} \\
& x_{i,j} \geq 0 \; \forall \, i, j &
\end{aligned}
$$

If we simply run this model, as is, the minimum cost plan would be to just do nothing! The cost would be zero. In reality, even though we are focused on costs in this application, there is an implied revenue and, therefore, profit (we hope!) that we aren't directly modeling. We are likely to instead be wanting to ship all of the product that we can at the lowest possible cost. More precisely, what we want to do is instead determine if the problem is supply limited or demand limited. This is a simple matter of comparing the net demand vs. the net supply and making sure that the lesser is satisfied completely.

TABLE 3.10 Demand and Supply Constrained Transportation Problems

If...	Then Situation is:	Source Constraints	Demand Constraints
$\sum_i S_i < \sum_j D_j$	Supply Constrained	$\sum_j x_{i,j} = S_i$	$\sum_i x_{i,j} \leq D_j$
$\sum_i S_i > \sum_j D_j$	Demand Constrained	$\sum_j x_{i,j} \leq S_i$	$\sum_i x_{i,j} = D_j$
$\sum_i S_i = \sum_j D_j$	Balanced	$\sum_j x_{i,j} = S_i$	$\sum_i x_{i,j} = D_j$

In the balanced situation, either source or demand constraints can be equalities.

Similarly, if we try to use equality constraints for both the supply and demand nodes but the supply and demand are not balanced, the LP will not be feasible.

In `ompr`, a double subscripted non-negative variable, $x_{i,j}$ can be defined easily as the following: `add_variable(x[i, j], type = "continuous", i = 1:10, j = 1:10, lb=0)`. Let's show how to set up a basic implementation of the transportation problem in `ompr`. This demonstrates the use of double subscripted variables. Also, it should be customized for a particular application of being demand or supply constrained, but it will give a good running head start.

```
NSupply <- 3    # 3 Supply nodes
NDest   <- 4    # 4 Destination nodes
snames <- list("MN", "PA", "WY")
dnames <- list("Atl", "Bos", "Chi", "Den")
Cost <- matrix (c(65, 36, 65, 57, 30, 70,
                  22, 30, 55, 42, 60, 42),
                nrow = NSupply,
                dimnames =list(snames,dnames))
S <- c(1000, 1500, 1500)

D <- c(700,  900, 1200, 900)
```

At this point, it may be a good idea to look at the arcs and see if you can expect which routes are likely to be heavily used and which are likely to be unused. You can also do this by looking at the matrix of transportation costs.

TABLE 3.11 Costs for Transportation Application

	Atl	Bos	Chi	Den
MN	65	57	22	42
PA	36	30	30	60
WY	65	70	55	42

Now, let's move on to implementing the model.

```
transportationmodel <- MIPModel()              |>
 add_variable(Vx[i, j], type = "continuous",
              i = 1:NSupply,
              j = 1:NDest, lb=0)               |>
 set_objective (sum_expr(Cost[i,j] * Vx[i,j] ,
                         i=1:NSupply,
                         j=1:NDest ), "min") |>
 add_constraint (sum_expr(Vx[i,j], i=1:NSupply)
                 >= D[j],
                 j=1:NDest)|>
 add_constraint (sum_expr(Vx[i,j], j=1:NDest)
                 <= S[i],
                 i=1:NSupply)

res.transp <-solve_model(
  transportationmodel, with_ROI(solver ="glpk"))
```

Let's confirm that our model solved correctly.

```
res.transp$status
```

```
## [1] "optimal"
```

Since it was solved to optimality, the optimal objective function value is meaningful and again we can use an inline code chunk of `R` to extract this value 1.259×10^5. The actual decision variable values of the solution are more interesting though. Let's first prepare these results by extracting them from the solution object and then applying our names to them.

```
Transp <- matrix(res.transp$solution,
                 nrow = NSupply,
                 dimnames =list(snames,dnames))
```

Now we can display the table of optimal transportation decisions from supply points to destinations cleanly.[1]

TABLE 3.12 Optimal Transportation Plan

	Atl	Bos	Chi	Den
MN	0	0	1000	0
PA	600	900	0	0
WY	100	0	200	900

3.4.3 Transshipment Models

A generalization of the transportation model is that of transshipment where some nodes are intermediate nodes that are neither pure sources or destinations but can have both inflow and outflow. We could extend our previous example with factories in Vietnam and Thailand. Note that the facilities in the middle (Minnesota, Pennsylvania, and Wyoming) are better described as distribution centers rather than warehouses.

In this case, the standard convention might be to work from left to right and index the set of factories by i, the set of distribution centers by j and the set of customers by k. We might also use $x_{i,j}$ to be the amount to ship from factory i to distribution center j and $y_{j,k}$ to be the amount to ship from distribution center j to customer k.

A common characteristic of this kind of problem is then that the inflow at each distribution center must equal the outflow. This means that for Minnesota, $(j = 1)$, we would would have a relationship of the following.

$$x_{1,1} + x_{2,1} = y_{1,1} + y_{1,2} + y_{1,3} + y_{1,4}$$

This lends itself to a summation.

$$\sum_{i=1}^{2} x_{i,1} = \sum_{k=1}^{4} y_{1,k}$$

[1]ompr 1.0 changed the order of the solution results. Adding the option `byrow=T` to the matrix command fixes this.

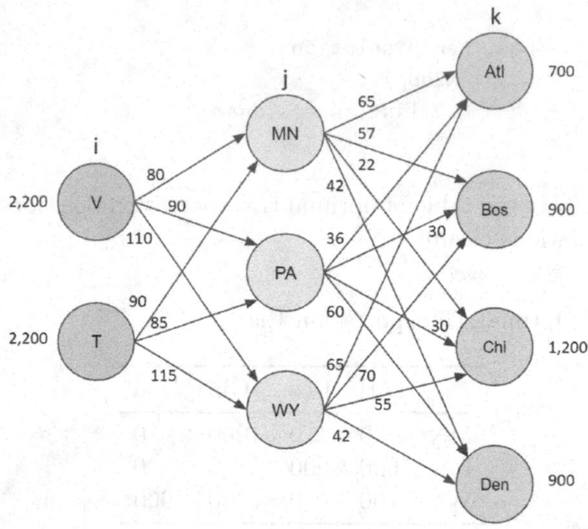

FIGURE 3.2 Transshipment Application.

We can then generalize this for every distribution center, j.

$$\sum_{i=1}^{2} x_{i,j} = \sum_{k=1}^{4} y_{j,k} \ \forall \ j$$

Now we could put everything together. We'll differentiate the costs to ship from supply and to destination now as $C^S_{i,j}$ and $C^D_{j,k}$, respectively.

$$\begin{aligned}
\text{Min } & \sum_i \sum_j C^S_{i,j} x_{i,j} + \sum_j \sum_k C^D_{j,k} y_{j,k} & \\
\text{s.t.: } & \sum_j y_{j,k} \leq D_k \ \forall \ k & \text{[Demand Limits]} \\
& \sum_j x_{i,j} \leq S_i \ \forall \ i & \text{[Supply Limits]} \\
& \sum_i x_{i,j} = \sum_k y_{j,k} \ \forall \ j & \text{[Inflow=Outflow]} \\
& x_{i,j}, y_{j,k} \geq 0 \ \forall \ i, j, k &
\end{aligned}$$

The directions of the supply and demand constraint inequalities must again be considered as to whether the problem is supply or demand constrained, lest we find an optimal solution of doing nothing.

This transshipment application can be further enhanced to allow for losses along arcs. In this case there might be affect amount that reaches the distribution center based on losses enroute. For example, we might use $L_{i,j}$ to capture this effect with $L_{1,2} = 0.9$ to indicate that only 90% of the material makes it from factory *1* to distribution center *2*. We would then modify the balance constraints to take the following form $\sum_i L_{i,j} \cdot x_{i,j} = \sum_k y_{j,k} \; \forall \; j$. Losses might also apply in the transportation to the end customer with similar modifications to the model.

Another common situation is to add capacity limits to each arc, perhaps both upper and lower bounds on capacity.

Real world applications applications often also require modeling a time dimension which leads us to issue production and inventory planning.

3.4.4 Production and Inventory Planning

A common application is production planning and inventory management planning over time. Let's assume a company has a manufacturing cost for product p in time period t of $C^M_{p,t}$. The cost to carry a unit of product p in inventory from period t to $t+1$ is $C^I_{p,t}$. Demand of $D_{p,t}$ must be met each period. The maximum production in any period can vary and is denoted as $M_{p,t}$. The beginning inventory for each product is B_p. The goal of the manager is to find a production plan that minimizes cost.

We can start by defining our decision variables. We will need a set of decision variables for the amount to produce of each product p in each period t. Let's define that as $x_{p,t}$.

Similarly, we will define a variable for the inventory at the end of each period, $y_{p,t}$.

The production cost is simply cost per product multiplied by the number of products produced or $\sum_p \sum_t C^M_{p,t} \cdot x_{p,t}$.

The inventory carrying cost follows the same structure, $\sum_p \sum_t C^I_{p,t} \cdot y_{p,t}$.

We can then combine them as $\sum_p \sum_t \left(C^M_{p,t} \cdot x_{p,t} + C^I_{p,t} \cdot y_{p,t}\right)$

At this point, it is important to look at the relationship between inventory, production, and sales.

Sometimes care must be taken for the "boundary conditions" of the first or last time period. In our case, we have a value defined the beginning inventory of B_p. For product 1, we would have $B_1 + x_{1,1} - D_{1,1} = y_{1,1}$. For product 2, it would be $B_2 + x_{2,1} - D_{2,1} = y_{2,1}$ and so on for other products. We would generalize this then as $B_p + x_{p,1} - D_{p,1} = y_{p,1}$.

For periods after the first ($t > 1$)) Essentially the ending inventory is equal to the beginning inventory plus the inflow (production) minus the outflow (demand). For product p in time period t, this then becomes $y_{p,t-1} + x_{p,t} - D_{p,t} = y_{p,t}$.

Of course we would need to ensure appropriate upper and lower bounds on the decision variables.

Let's wrap it all together in the follow formulation.

$$
\begin{aligned}
\text{Min} & \sum_p \sum_t \left(C^M_{p,t} \cdot x_{p,t} + C^I_{p,t} \cdot y_{p,t} \right) & \\
\text{s.t.:} \; & B_p + x_{p,1} - D_{p,1} = y_{p,1} \; \forall \; p & \text{[Period 1 Inventory]} \\
& y_{p,t-1} + x_{p,t} - D_{p,t} = y_{p,t} \; \forall \; p, \; t > 1 & \text{[Following Periods]} \\
& 0 \leq x_{p,t} \leq M_{p,t} \; \forall \; p, t & \text{[Production Limits]} \\
& x_{p,t} \; y_{p,t} \geq 0 \; \forall \; p, t & \text{[Non-negativity]}
\end{aligned}
$$

This model illustrates the basic dynamics of relating production, inventory, and demand. Each application often requires tailoring to address specific needs. For example:

- Some companies require safety stocks levels to be maintained which could be done by setting minimum inventory levels.
- Sometimes all the demand cannot be met in a period – this can be modeled by separating sales from demand with a separate variable for sales in each period.
- Allowing backlogged demand that can be met in the following period – again modeled with an additional variable.

3.4.5 Standard Form

Any linear program with inequality constraints can be converted into what is referred to as standard form. First, all strictly numerical terms are collected or moved to the right-hand side and all variables are on the left-hand side.

It makes little difference as to whether the objective function is a *min* or a *max* function since a min objective function can be converted to a max objective function by multiplying everything by a negative one. The converse is also true.

The last step is where all of the inequalities are replaced by strict equality relations. The conversion of inequalities to equalities warrants a little further explanation. This is done by introducing a new, non-negative "slack" variable for each inequality. If the inequality is a $\leq$, then the slack variable can be thought of as filling the left-hand side to make it equal to the right-hand side so the slack variable is added to the left-hand side. If the inequality is $\geq$, then the slack variable is the amount that must be absorbed from the left-hand side to make it equal the right-hand side.

Let's demonstrate with a two variable example.

$$\begin{aligned} \max\ & 2 \cdot x + 3 \cdot y \\ \text{s.t.:}\ & 10 \cdot x + 2 \cdot y \leq 110 \\ & 2 \cdot x + 10 \cdot y \geq 50 \\ & 2 \cdot x = 4 \cdot y \\ & x,\ y \geq 0 \end{aligned}$$

This can be then transformed to standard form by rearranging terms in the third constraint and adding two non-negative slack variables to the first two constraints.

$$\begin{aligned} \max\ & 2 \cdot x + 3 \cdot y \\ \text{s.t.:}\ & 10 \cdot x + 2 \cdot y + s_1 = 110 \\ & 2 \cdot x + 10 \cdot y - s_2 = 50 \\ & 2 \cdot x - 4 \cdot y = 0 \\ & x,\ y,\ s_1,\ s_2 \geq 0 \end{aligned}$$

The standard form now consists of a system of equations, generally with far more variables (both regular and slack) than equations. The simplex algorithm makes use of the fact that a system of equations and unknowns can be solved efficiently. The Simplex algorithm solves for as many variables (termed basic variables) as there are equations and sets the remaining variables to zero (non-basic variables). It then systematically swaps a variable from the set of basic variables and non-basic variables until it can no longer find an improvement. The actual algorithm for the Simplex algorithm is beyond the scope of what we will cover in this book.
Let's examine the solution to our example though.

```
StdModel <- MIPModel() |>
 # Avoid name space conflicts V prefix for ompr variables.

 add_variable(Vx, type = "continuous", lb = 0)  |>
 add_variable(Vy, type = "continuous", lb = 0)  |>
 add_variable(Vs1, type = "continuous", lb = 0) |>
 add_variable(Vs2, type = "continuous", lb = 0) |>

 set_objective(2*Vx + 3*Vy,"max")               |>

 add_constraint(10*Vx + 2*Vy + Vs1==110)        |>
 add_constraint(2*Vx + 10*Vy - Vs2==50)         |>
 add_constraint(2*Vx - 4*Vy == 0)

  resStd <- solve_model(StdModel,
                        with_ROI(solver = "glpk"))
  resStd
```

```
## Status: optimal
## Objective value: 35
```

```
x  <- get_solution (resStd , Vx)
y  <- get_solution (resStd , Vy)
s1 <- get_solution (resStd , Vs1)
s2 <- get_solution (resStd , Vs2)

base_case_res           <- cbind(x, y, s1, s2)
rownames(base_case_res) <- "Optimal Values"
kbl(base_case_res, booktabs=T,
   caption="Solution to Standard Form")|>
  kable_styling(latex_options = "hold_position")
```

Notice that we had three equations and four variables (both regular and slack variables.) This meant that at every iteration of the simplex algorithm, it would set one variable equal to zero and solve for the other three variables because solving three equations with three unknowns is very easy. This is consistent with our solution, where s_1 is zero and the other three variables all had values.

TABLE 3.13 Solution to Standard Form

	x	y	s1	s2
Optimal Values	10	5	0	20

3.5 Vector and Matrix Forms of LPs

One of the benefits of using an algebraic modeling approach for linear programming such as `ompr` is that it provides a direct mapping from the mathematical model of the application to the implementation. The actual solver used such as `glpk`, `symphony`, or `lpSolve` take the linear program in a different form though and `ompr` processes it into a form they can handle. Technically, the `glpk` and `symphony` are made available on CRAN as `Rglpk` and `Rsymphony` to distinguish them from other implementations of their optimization engines.

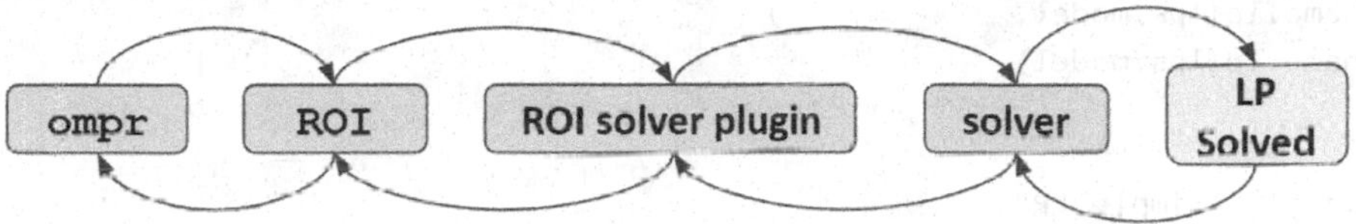

FIGURE 3.3 Relationship between ompr and solvers.

In general, the solver thinks of the problem in terms of a vector of coefficients for the objective function, C, a vector of right-hand side constraint values, B, a matrix of data, A, and a vector of variables, x. The result is that any linear program can be thought of in a form to the right.

$$\begin{aligned}\max\ & C \cdot x \\ \text{s.t.: } & A \cdot x \leq B \\ & x \geq 0\end{aligned}$$

Let's examine building our first two variable LP model in this way using the `lpsolveAPI` package.

$$\begin{aligned}
\text{Max } & 7 \cdot Ants + 12 \cdot Bats \\
\text{s.t.: } & 1 \cdot Ants + 4 \cdot Bats \leq 800 \\
& 3 \cdot Ants + 6 \cdot Bats \leq 900 \\
& 2 \cdot Ants + 2 \cdot Bats \leq 480 \\
& 2 \cdot Ants + 10 \cdot Bats \leq 1200 \\
& Ants,\ Bats \geq 0
\end{aligned}$$

Now let's implement the model the model using `lpSolveAPI`.

```
library(lpSolveAPI)
lps.model <- make.lp(0, 2) # Make empty 2 var model
add.constraint(lps.model, c(1,4), "<=", 800)
add.constraint(lps.model, c(3,6), "<=", 900)
add.constraint(lps.model, c(2,2), "<=", 480)
add.constraint(lps.model, c(2,10), "<=", 1200)
set.objfn(lps.model, c(-7,-12))
name.lp(lps.model, "Simple LP")
name.lp(lps.model)
```

```
## [1] "Simple LP"
```

```
# plot.lpExtPtr(lps.model)
solve(lps.model)
```

```
## [1] 0
```

```
get.primal.solution(lps.model, orig=TRUE)
```

```
## [1] 420 900 480 960 180  60
```

While the graph may not be as elegant our earlier versions, it shows how the problem can be visualized.

Earlier in the chapter, we created data structures for the generalized version of the optimization model. Recall that earlier in this chapter we created a `Resources` matrix of the resources used by each product – this corresponds to our A matrix. We also created a single column matrix, `Available`, for the amount of each resource available, which is our B vector. The `Profit` single row matrix serves as our C vector. The last item that we need to do is to specify a direction for each inequality.

We could rewrite our formulation then as the following.

$$\begin{aligned} \max\ & Profit \cdot x \\ \text{s.t. } & Resources \cdot x \leq Available \\ & x \geq 0 \end{aligned}$$

The `Rglpk` package can then be accessed directly.

FIGURE 3.4 Relationship between ompr and Rglpk.

```
library (Rglpk)     # Load Rglpk package

dir2<- c( "<=", "<=", "<=", "<=", "<=")
                    # Set direction of each inequality

res2 <- Rglpk_solve_LP(obj=Profit,     # C vector
                       mat=Resources, # A matrix
                       dir=dir2,  # Constraint Inequalities
                       Available, # B vector
                       max=TRUE)  # False does minimization
  # Uses default bounds on all variables:
  #    LB = 0, UB=infinity (just non-negativity)
  #    Defaults to continuous variables
```

Notice that calling the solver directly such as in this way using `Rglpk` does not require naming decision variables. A decision variable is created for each column of the matrix. Directly accessing the solver allows for all the same features that we use through `ompr` such as placing bounds on the variables, changing to a minimization objective function, and more such as advanced solving options. Other LP solving engines available in `R` such as `RSymphony` and `lpSolve` have similar syntax for solving. Let's look at the results. As usual, we start by examining the status.

```
res2$status
```

```
## [1] 0
```

The status returned by Rglpk is zero. This may sound bad but the package's help file indicates that returning a status value of zero means that no problems occurred, and it found an optimal solution. We can then proceed to examine the objective function value.

```
res2$optimum
```

```
## [1] 4800
```

This objective function value may look familiar. Now let's look at the solution in terms of optimal decision variable values.

```
res2$solution
```

```
## [1] 240   0   0   0
```

Most of these LP solving engines can also allow models to be built up incrementally as well by creating an empty model, then adding an objective function and adding constraints, bounds on variables as separate lines of R code.

While working with the solver directly has the benefit of avoiding ambiguities such as the difficulty of parsing solution status correctly, creating a single A matrix for optimization problems with multiple sets of decision variables and decision variables that may have 3 or more subscripts can be very tricky, hard to maintain, and even harder to debug. Also, the results require just as much or more care in unpacking as the solution variable values are returned as a single vector, x containing all the separate decision variables. Think of having three triple subscripted variables, u, v, and y with each subscript having 12 values. The resulting x vector will have $3 \cdot 12^3 = 5184$ elements, each of which must always be considered in exactly the same order for each constraint or building of the A matrix.

At this point, it may be apparent that a chief benefit of ompr is that allows the modeler to deal with the model at a higher level and forces the computer to handle the detailed accounting translating the model into a lower level solvable structure and then translating the solution back into the original terms. These conversions are analogous to using a high-level computer language versus a low level assembly code. A talented programmer can write a program in assembly code to do anything that could be done in a high-level language but the cost in terms of time and effort is typically prohibitive.

The investment of time and effort to use a lower level, more direct access to the solver engine may be warranted when developing a package that does optimization. For example, packages for doing Data Envelopment Analysis require doing many linear programs. These are typically done using the lower level engines such as lpSolveAPI, `Rglpk`, `lpSolveAPI`, or other tools rather than `ompr` but the platform decision should be made carefully. A more indepth example of building optimization models using lower level access calls is in *DEA Using R*.

Since each low level solver may have different choices made in terms of the name or format of parameters to be passed for the LP as well as a different naming convention of elements being passed back, a translation layer has been developed, `ROI` which stands for the R Optimization Interface. There is the core `ROI` package and a specific `ROI` interface for each LP solver such as `ROI.plugin.glpk`. `ROI` can then be used to simplify switching between LP solvers. This is demonstrated in Chapter 8.

3.6 Exercises

Exercise 3.1 (Transportation). Four manufacturing plants are supplying material for distributors in four regions. The four supply plants are located in Chicago, Beaverton, Eugene, and Dallas. The four distributors are in PDX (Portland), SEA (Seattle), MSP (Minneapolis), and ATL (Atlanta). Each manufacturing plant has a maximum amount that they can produce. For example, Chicago can produce at most 500. Similarly, the PDX region can handle at most 700 units. The cost to transport from Dallas to MSP is three times as high as the cost from Dallas to Atlanta. The following table displays the transportation cost between all the cities.

TABLE 3.14 Transportation Cost between Cities

Node	PDX	SEA	MSP	ATL	**Supply**
Chicago	20	21	8	12	**500**
Beaverton	6	7	18	24	**500**
Eugene	8	10	22	28	**500**

Node	PDX	SEA	MSP	ATL	**Supply**
Dallas	16	26	15	5	**600**
Capacity	**700**	**500**	**500**	**600**	

Formulate an explicit model for the above application that solves this transportation problem to find the lowest cost way of transporting as much as product as we can to distributors. Hint: You might choose to define variables based on the first letter of source and destination so *XCP* is the amount to ship from Chicago to PDX.

Implement and solve the model using `ompr`. Be sure to interpret and discuss the solution as to why it makes it sense.

Exercise 3.2 (Generalized Transportation Model). Formulate a generalized model for the above application that solves this transportation problem to find the lowest cost way of transporting as much product as we can to distributors.

Hint: Feel free to use my LaTeX formulation for the general transportation model and make change(s) to reflect your case.

Implement and solve the model using `ompr`. Be sure to discuss the solution as to why it makes sense.

Exercise 3.3 (Convert LP to Standard Form). Convert the three variable LP represented in Table 3.13 into standard form.

$$
\begin{aligned}
\text{Max:} & \\
& Profit = 20 \cdot Ants + 10 \cdot Bats + 16 \cdot Cats \\
\text{S.t.} & \\
& 6 \cdot Ants + 3 \cdot Bats + 4 \cdot Cats \leq 2000 \\
& 8 \cdot Ants + 4 \cdot Bats + 4 \cdot Cats \leq 2000 \\
& 6 \cdot Ants + 3 \cdot Bats + 8 \cdot Cats \leq 1440 \\
& 40 \cdot Ants + 20 \cdot Bats + 16 \cdot Cats \leq 9600 \\
& Cats \leq 200 \\
& Ants, Bats, Cats \geq 0
\end{aligned}
$$

Exercise 3.4 (Solve Standard Form). Implement and solve the standard form of the LP using R. Be sure to interpret the solution and discuss how it compares to the solution from the original model.

Exercise 3.5 (Adding a Constraint). Add a new constraint to model in Exercise 3.4 such that number of Bats are double than the number of Ants. Be sure to interpret the solution and discuss how it compares to the solution from the original model.

Exercise 3.6 (Convert Generalized LP to Standard Form). Convert the following generalized production planning LP into standard form.
Hint: define a set of variables, s_j to reflect these changes and add it to the following formulation.

$$\begin{aligned} \text{Maximize } & \sum_{i=1}^{3} P_i x_i \\ \text{subject to } & \sum_{i=1}^{3} R_{i,j} x_i \leq A_j \; \forall \; j \\ & x_i \geq 0 \; \forall \; i \end{aligned}$$

Exercise 3.7 (Staff Scheduling). The need for nurses has increased rapidly during the Covid pandemic. Specific Hospital has determined their need for nurses throughout the day. They expect the following numbers of nurses to be needed over the typical 24 hour period. The nurses work eight hour shifts, starting at each 4 hour time period specified in the table.

```
hph_needs <- matrix (c(50, 140, 260, 140, 100, 80), ncol=1,
    dimnames = c(list(c("midnight - 4 AM",
                        "4 AM - 8 AM",
                        "8 AM - noon",
                        "noon - 4 PM",
                        "4 PM - 8 PM",
                        "8 PM - midnight")),
                 list(c("Required Nurses"))))
kbl (hph_needs, booktabs=T, caption="Staff Scheduling") |>
  kableExtra::kable_styling(latex_options = "hold_position")
```

a. Create an appropriate optimization model to minimize the number of nurses needed to cover the requirements.
b. Implement and solve the model in R.
c. Discuss and explain your solution.

Exercise 3.8 (Reallocating personnel at Nikola). Due to the semiconductor shortage affecting the automotive industry, the production of vehicles has slowed down. The lack of a key control chip has brought the Engine shop of the manufacturing plant of automotive company Nikola to a complete halt.

TABLE 3.15 Staff Scheduling

	Required Nurses
midnight - 4 AM	50
4 AM - 8 AM	140
8 AM - noon	260
noon - 4 PM	140
4 PM - 8 PM	100
8 PM - midnight	80

The company has decided to use the personnel from the Engine shop in other shops in the Press shop and the Paint shop, split equally. Also, it has decided to store the assembled vehicles in the plant until Engine Shop is functional again. The maximum capacity of the storage of cars is 10000. Current manpower and production information is as given in the table.

Process Area	Alpha	Beta	Gamma	Theta	Available labor hours
Profit	\$4000	\$3000	\$3600	\$5000	
Press Shop	1	3	2	2	600
Weld Shop	3	4	2	2	1200
Paint Shop	2	3	1	2	760
Engine Shop	3	5	2	4	700
Assembly Shop	2	0	2	0	1000

Make required change to the given table and Create an optimization model to find a production plan with max profit for all four car models – Alpha, Beta, Gamma, and Theta. Make sure to consider the max storage limit of 10000 cars.

4

Sensitivity Analysis

We can get a lot more than just the objective function value and the decision variable values from a solved linear program. In particular, we can potentially explore the impact of changes in constrained resources, changes in the objective function, forced changes in decision variables, and the introduction of additional decision variables.

4.1 Base Case

To demonstrate this, let's revisit our explicit model of production planning. We will use the explicit version for the sake of clarity and simplicity but the same techniques could be used for the generalized model or other linear programs.

Recall the three variable explicit production planning problem from Chapter 2.

$$\begin{aligned} \text{Max } & 7 \cdot Ants + 12 \cdot Bats + 5 \cdot Cats \\ \text{s.t.: } & 1 \cdot Ants + 4 \cdot Bats + 2 \cdot Cats \leq 800 \\ & 3 \cdot Ants + 6 \cdot Bats + 2 \cdot Cats \leq 900 \\ & 2 \cdot Ants + 2 \cdot Bats + 1 \cdot Cats \leq 480 \\ & 2 \cdot Ants + 10 \cdot Bats + 2 \cdot Cats \leq 1200 \\ & Ants, Bats, Cats \geq 0 \end{aligned}$$

The implementation that we did earlier for production planning was straightforward.

```
Base3VarModel <- MIPModel() |>
  add_variable(Ants, type = "continuous", lb = 0) |>
```

DOI: 10.1201/9781003051251-4

```
  add_variable(Bats, type = "continuous", lb = 0) |>
  add_variable(Cats, type = "continuous", lb = 0) |>

  set_objective(7*Ants + 12*Bats + 5*Cats,"max")  |>

  add_constraint(1*Ants + 4*Bats + 2*Cats<=800)   |>
  add_constraint(3*Ants + 6*Bats + 2*Cats<=900)   |>
  add_constraint(2*Ants + 2*Bats + 1*Cats<=480)   |>
  add_constraint(2*Ants + 10*Bats + 2*Cats<=1200) |>
  solve_model(with_ROI(solver="glpk"))

base_case_res <- cbind(objective_value(Base3VarModel),
                       get_solution(Base3VarModel, Ants),
                       get_solution(Base3VarModel, Bats),
                       get_solution(Base3VarModel, Cats))
colnames(base_case_res)<-list("Profit", "Ants",
                              "Bats", "Cats")
rownames(base_case_res)<-list("Base Case")
kbl(base_case_res, booktabs=T,
    caption="Base Case Production Plan") |>
  kable_styling(latex_options = "hold_position")
```

TABLE 4.1 Base Case Production Plan

	Profit	Ants	Bats	Cats
Base Case	2225	50	0	375

In the base case, we are producing Ants and Cats but not Bats to generate a total profit of $2225.

4.2 Shadow Prices

4.2.1 Extraction and Interpretation

There are many resources, some are fully used, while some are not fully utilized. How do we prioritize the importance of each resource? For example, if the factory manager could add a worker to one department, which should it be? Conversely, if an outside task came up where should she draw the time from?

We could modify the model and rerun. For complex situations, this may be necessary. On the other hand, we could also use sensitivity analysis to explore the relative value of the resources.

Let's begin by examining the row duals, also known as shadow prices.

```
rduals1 <-as.matrix(get_row_duals(Base3VarModel))
dimnames(rduals1)<-list(c("Machining", "Assembly",
                          "Testing", "Sensors"),
                        c("Row Duals"))
kbl(format(rduals1,digits=4), booktabs=T,
    caption="Shadow Prices of Constrained Resources") |>
  kable_styling(latex_options = "hold_position")
```

TABLE 4.2 Shadow Prices of Constrained Resources

	Row Duals
Machining	0.25
Assembly	2.25
Testing	0.00
Sensors	0.00

The row duals or shadow prices for testing is zero. This means that the marginal value of one additional hour of testing labor time is \$0. This makes sense if you examine the amount of testing time used and realize that the company is not using all of the 480 hours available. Therefore adding more testing hours certainly can't help improve the production plan.

On the other hand, all of the assembly time (resource) is used in the optimal solution. The shadow price of an hour of assembly time is \$2.25. This means that for every hour of additional assembly time within certain limits, the objective function will increase by \$2.25 also.

All of the 900 hours of labor available in the assembly center are consumed by the optimal production plan. Increasing the assembly hours available may allow the company to change the production plan and increase the profit. While you could rerun the model with increased assembly hours to determine the new optimal production plan but if you only want to know the change in the optimal objective function value, you can determine that from the shadow price of the assembly constraint. Each additional hour (within a certain range) will increase the profit by \$2.25.

This potential for increased profit can't be achieved by simply increasing the resource – it requires a modified production plan to utilize this increased resource. To find the production plan that creates this increased profit, let's solve a modified linear program.

4.2.2 Example of Adding an Hour to Assembly

Let's test the numerical results from the Shadow Price table by adding an hour of labor to the assembly department. The model is represented in the following formulation.

$$
\begin{aligned}
\text{Max: } & 7 \cdot Ants + 12 \cdot Bats + 5 \cdot Cats \\
\text{s.t. } & 1 \cdot Ants + 4 \cdot Bats + 2 \cdot Cats \leq 800 \\
& 3 \cdot Ants + 6 \cdot Bats + 2 \cdot Cats \leq 901 \\
& 2 \cdot Ants + 2 \cdot Bats + 1 \cdot Cats \leq 480 \\
& 2 \cdot Ants + 10 \cdot Bats + 2 \cdot Cats \leq 1200 \\
& Ants,\ Bats,\ Cats \geq 0
\end{aligned}
$$

The code implements the revised model.

```
IncAssemHrs <- MIPModel() |>
  add_variable(Ants, type = "continuous", lb = 0)  |>
  add_variable(Bats, type = "continuous", lb = 0)  |>
  add_variable(Cats, type = "continuous", lb = 0)  |>

  set_objective(7*Ants + 12*Bats + 5*Cats,"max")  |>

  add_constraint(1*Ants + 4*Bats + 2*Cats<= 800)   |>
  add_constraint(3*Ants + 6*Bats + 2*Cats<= 901)   |>
  add_constraint(2*Ants + 2*Bats + 1*Cats<= 480)   |>
  add_constraint(2*Ants + 10*Bats + 2*Cats<= 1200) |>

  solve_model(with_ROI(solver = "glpk"))

inc_assem_res  <- cbind(objective_value(IncAssemHrs),
                      get_solution(IncAssemHrs, Ants),
                      get_solution(IncAssemHrs, Bats),
                      get_solution(IncAssemHrs, Cats))
colnames(inc_assem_res)<-list("Profit",
                              "Ants",
                              "Bats",
                              "Cats")
```

```
rownames(inc_assem_res)<-list("+1 Assembly Hr")
temp1 <- rbind(base_case_res, inc_assem_res)
kbl(temp1, booktabs=T,
   caption="Production Plan with One Additional Assembly Hour") |>
  kable_styling(latex_options = "hold_position")
```

TABLE 4.3 Production Plan with One Additional Assembly Hour

	Profit	Ants	Bats	Cats
Base Case	2225.00	50.0	0	375.00
+1 Assembly Hr	2227.25	50.5	0	374.75

These results confirmed that adding one hour of Testing time results in a *new* production plan that generates an increased profit of \$2.25, exactly as expected.

While it is easy to look at an individual resource when we are looking at problems with only a couple of constraints, the shadow prices can be very helpful in larger problems with dozens, hundreds, or thousands of resources where questions might come up of trying to evaluate or prioritize limited resources.

4.2.3 Shadow Prices of Underutilized Resources

The shadow price on sensors is zero (as was also the case for testing hours). This means that even a large increase in the number of sensors would not affect the maximum profit or the optimal production plan. Essentially there is plenty of sensors available, having more would not allow a better profit plan to be possible. Let's confirm this as well with a numerical example by adding *10,000* more sensors.

$$
\begin{split}
\text{Max: } & 7 \cdot Ants + 12 \cdot Bats + 5 \cdot Cats \\
\text{s.t. } & 1 \cdot Ants + 4 \cdot Bats + 2 \cdot Cats \leq 800 \\
& 3 \cdot Ants + 6 \cdot Bats + 2 \cdot Cats \leq 900 \\
& 2 \cdot Ants + 2 \cdot Bats + 1 \cdot Cats \leq 480 \\
& 2 \cdot Ants + 10 \cdot Bats + 2 \cdot Cats \leq 11200 \\
& Ants, \; Bats, \; Cats \geq 0
\end{split}
$$

In the same way that it was done for adjusting assembly hours earlier, the next chunk shows how to examine the case of a huge increase in the number of sensors.

```
IncSensor<- MIPModel() |>
  add_variable(Ants, type = "continuous", lb = 0) |>
  add_variable(Bats, type = "continuous", lb = 0) |>
  add_variable(Cats, type = "continuous", lb = 0) |>

  set_objective(7*Ants + 12*Bats + 5*Cats,"max")  |>

  add_constraint(1*Ants+4*Bats + 2*Cats<= 800)     |>
  add_constraint(3*Ants+6*Bats + 2*Cats<=900)      |>
  add_constraint(2*Ants+2*Bats+1*Cats<=480)        |>
  add_constraint(2*Ants+10*Bats+2*Cats<=11200)     |>
  solve_model(with_ROI(solver = "glpk"))

inc_sensor_res  <- cbind(objective_value(IncSensor),
                     get_solution (IncSensor, Ants),
                     get_solution (IncSensor, Bats),
                     get_solution (IncSensor, Cats))
colnames(inc_sensor_res)<-list("Profit",
                               "Ants",
                               "Bats",
                               "Cats")
rownames(inc_sensor_res)<-list("Increased Sensors")
temp2 <- rbind(base_case_res, inc_assem_res, inc_sensor_res)
```

TABLE 4.4 Production Plan with 10,000 More Sensors

	Profit	Ants	Bats	Cats
Base Case	2225.00	50.0	0	375.00
+1 Assembly Hr	2227.25	50.5	0	374.75
Increased Sensors	2225.00	50.0	0	375.00

Even this massive increase of sensors does not result in any increase in profit or change the production plan.

4.3 Reduced Costs of Variables

Next, we move on to the *reduced costs* of variables. The reduced cost for a variable is the per unit marginal profit (objective function coefficient) minus the per unit value (in terms of shadow prices) of the resources used by a unit in production. The reduced costs is also often referred to as the column duals. The concept and use of reduced costs frequently may require rereading several times. The mathematical details rely on the structure of linear programming and the Simplex method. We won't go into detail on the origin of this mathematically in detail here.

4.3.1 Reduced Cost of Ants

Let's start by examining the Ants. The reduced cost for *Ants* is the per unit marginal profit minus the per unit value (in terms of shadow prices) of the resources used by a unit in production.

Let's extract the reduced costs from the results just as we did for the shadow prices. Note that the reduced costs are referred to as column duals in `ompr`.

```
cduals1 <-as.matrix(get_column_duals(Base3VarModel) )
dimnames(cduals1)<-list(c("Ants", "Bats", "Cats"),
                        c("Column Duals"))
```

TABLE 4.5 Reduced Costs of Variables

	Column Duals
Ants	0.0
Bats	−2.5
Cats	0.0

These results are interesting. The reduced costs of variables that are between simple upper and lower bounds will be zero. Again, the reduced cost of a variable is the difference between the value of the resources consumed by the product and the value of the product in the objective function. All of our product's variables have simple lower bounds of zero and no upper bounds. Ants and Cats have zero reduced cost while Bats have a negative reduced cost. Let's see if this is consistent with the interpretation of reduced costs.

Let's start by examining the shadow prices of the resources along with the number of each resource used in the production of a single Ant.

```
ant_res_used<-cbind(rduals1,c(1,3,2,2))
colnames(ant_res_used)<-c("Row Duals", "Resources Used")
ant_res_used <- rbind(ant_res_used,
```

TABLE 4.6 Resources Used by an Ant and Shadow Prices of Resources

	Row Duals	Resources Used
Machining	0.25	1
Assembly	2.25	3
Testing	0	2
Sensors	0	2
	TOTAL	7

```
                 c("TOTAL",t(rduals1)%*%c(1,3,2,2)))
kbl(format(ant_res_used, digits=4), booktabs=T,
    caption="Resources Used by an Ant and shadow prices of resources")
```

Taking the product in each row and adding them up will give the marginal value of the resources consumed by an incremental change in production of Ants. In this case, the marginal value is $1 \cdot 0.25 + 3 \cdot 2.25 + 2 \cdot 0 + 2 \cdot 0 = 7$. Since the profit per Ant (from the objective function coefficient on Ants) is also \$7, they are in balance and the difference between them is zero, which is why the reduced cost for Ants is zero. This can be thought of saying that the marginal benefit is equal to the marginal cost of a very small forced change in the value of the Ants variable. The value of the resources used in producing an Ant equals that of the profit of an Ant at the optimal solution.

We can repeat the same process and calculations for Cats

```
cat_res_used<-cbind(rduals1,c(2,2,1,2))
colnames(cat_res_used)<-c("Row Duals", "Resources Used")
cat_res_used <- rbind(cat_res_used,
                 c("TOTAL",t(rduals1)%*%c(2,2,1,2)))
kbl(format(cat_res_used, digits=4), booktabs=T,
    caption="Resources Used by a Cat and their Shadow Prices") |>
  kable_styling(latex_options = "hold_position")
```

In this case, using the columns from Table 4.6 we can calculate the marginal value as $2 \cdot 0.25 + 2 \cdot 2.25 + 1 \cdot 0 + 2 \cdot 0 = 5$. Since the profit per Cat (from the objective function coefficient on Cats) is also \$5, they are in balance and the difference between them is zero, which is why the reduced cost for the *Cats* variable is also zero.

TABLE 4.7 Resources Used by a Cat and Their Shadow Prices

	Row Duals	Resources Used
Machining	0.25	2
Assembly	2.25	2
Testing	0	1
Sensors	0	2
	TOTAL	5

4.3.2 Reduced Price of Bats

The situation is more interesting for Bats. The production plan does not call for producing any Bats. This means that it is sitting right at the lower bound of zero Bats. This hints that perhaps we would like to make even less than zero Bats if we could and that being *forced* to make even one Bat would be costly. A reduced cost of $\$-2.5$ means that the opportunity cost of resources used in producing a Bat is \$2.5 more than the profit of producing a single Bat In other words, we would expect that if we force the production of a single Bat, the over all production plan's profit will go down by \$2.5.

Let's go ahead and test this interpretation by again looking at the value of the resources consumed in the production of a Bat and the amount of each resource used.

```
bat_res_used<-cbind(rduals1,c(4,6,2,10))
colnames(bat_res_used)<-c("Row Duals", "Resources Used")
bat_res_used <- rbind(bat_res_used,
                      c("TOTAL",t(rduals1)%*%c(4,6,2,10)))
```

TABLE 4.8 Resources Used by a Bat and Their Shadow Prices

	Row Duals	Resources Used
Machining	0.25	4
Assembly	2.25	6
Testing	0	2
Sensors	0	10
	TOTAL	14.5

Notice that the values based on shadow prices of the resources used by a Bat are $4 \cdot 0.25 + 6 \cdot 2.25 + 2 \cdot 0 + 10 \cdot 0 = 14.5$. Alas, the profit for each Bat is just \$12 which means that forcing the production of a single Bat will decrease the production plan's profit by $\$12 - 14.5 = -2.5$. In other words, the impact on the objective function is $\$ - 2.5$, which is the same as the reduced price of Bats.

Now let's test it. We will modify the formulation to set a lower bound on the number of Bats to be 1. Note that we do this in case by setting the `lb` option in the `add_variable` to be 1. Also, if we had a demand constraint for Bats, we could also be accommodated this by setting the upper bound (`ub`).

```
Bat1Model <- MIPModel() |>
  add_variable(Ants, type = "continuous", lb = 0) |>
  add_variable(Bats, type = "continuous", lb = 1) |>
  add_variable(Cats, type = "continuous", lb = 0) |>

  set_objective(7*Ants + 12*Bats + 5*Cats,"max")  |>

  add_constraint(1*Ants+4*Bats+2*Cats<=800)       |>
  add_constraint(3*Ants+6*Bats+2*Cats<=900)       |>
  add_constraint(2*Ants+2*Bats+1*Cats<=480)       |>
  add_constraint(2*Ants+10*Bats+2*Cats<=1200)     |>

  solve_model(with_ROI(solver = "glpk"))

  Bat1_case_res <- cbind(objective_value(Bat1Model),
                         get_solution (Bat1Model, Ants),
                         get_solution (Bat1Model, Bats),
                         get_solution (Bat1Model, Cats))
```

Let's compare the results for the new production plan and the original base case looking at the Table 4.9.

```
rownames(base_case_res) <- "Base Case"
rownames(Bat1_case_res) <- "Force one Bat"
temp3 <-rbind(base_case_res,Bat1_case_res)
kbl(temp3, booktabs=T,
    caption="Impact of a Forced Change in Bats") |>
  kable_styling(latex_options = "hold_position")
```

As we expected, the forced change of making one additional Bat resulted in a decrease of the overall profit from \$2225 to \$2222.5. This occurred because

TABLE 4.9 Impact of a Forced Change in Bats

	Profit	Ants	Bats	Cats
Base Case	2225.0	50	0	375.0
Force one Bat	2222.5	49	1	373.5

making one Bat meant that we had fewer of the precious, limited resources, decreasing overall profit due to limiting our possible production of Ants and Cats.

This meant that the number of Ants and Cats were changed resulting in a lower total profit even though a Bat is on its own profitable.

Another way to view the reduced costs for Bats is to think of it as an opportunity cost for not being able to further change the production of Bats due the simple bound. Think of it as a cost for being pinned at the simple upper or lower bound (often 0). A non-zero reduced cost means that the optimizer would like to relax or loosen the lower bound and the value is how much better the objective would be with a unit change of one relaxing the bound. For example, in our case, with bats being pinned at zero at the optimal solution and a reduced cost of bats of −2.5, this means that if we relaxed the non-negativity from $Bats \geq 0$ to instead be $Bats \geq -1$ and re-solve, then the objective function would improve by 2.5. In other words, relaxing the simple lower bound would change the production plan to enable profit to increase by 2.5.

4.4 Using Sensitivity Analysis to Evaluate a New Product

Let's consider a design proposal for a Dog drone. The Dog has a projected profit of $20 each and uses 8 hours of machining time, 12 of assembly, and 4 of testing. Each dog drone also uses 4 sensors.

```
dog_res_used<-cbind(rduals1,c(8,12,4,4))
colnames(dog_res_used)<-c("Row Duals", "Resources Used")
dog_res_used <- rbind(dog_res_used,
                    c("TOTAL",t(rduals1)%*%c(8,12,4,4)))
```

TABLE 4.10 Resources Used by a Dog and Their Shadow Prices

	Row Duals	Resources Used
Machining	0.25	8
Assembly	2.25	12
Testing	0	4
Sensors	0	4
	TOTAL	29

Even without adding it to the model, we can check to see if it is worthwhile to consider seriously. The opportunity cost of producing one Dog drone would result in the $\$20 - (8 \cdot 0.25 + 12 \cdot 2.25 + 4 \cdot 0 + 4 \cdot 0) = -9.0$. In other words, even though a Dog drone has a much higher profit than the other products, producing one would cost the company nine dollars of overall profit in terms of the opportunity cost in lost profit of other production.
We could interpret this as a simple hard stop on the decision to produce Dogs but we could go one step further by setting a target for redesigning the product or its production process. If the assembly time could be reduced by four hours to just eight hours, then the value of the resources consumed would be equal to the profit, and we would be indifferent to producing some Dog drones.

4.5 Exercises

Exercise 4.1 (Adding Eels). Your company has extended production to allow for producing aquatic Eels and is now including a finishing department that primes and paints the drones.

TABLE 4.11 Adding Eels

Characteristic	Ants	Bats	Cats	Eels	Available
Profit	\$7	\$12	\$5	\$22	
Machining	1	4	2	4	800
Assembly	3	6	2	8	900
Testing	2	2	1	25	480
Sensor	2	10	2	16	1200
Painting	1	1	1	12	500

a. Use R Markdown to create your own description of the model.

b. Extend the R Markdown to show your LP Model. Be sure to define models.
c. Solve the model in R.
d. Interpret and discuss the model in R Markdown.
e. Examine and reflect upon the reduced costs and shadow prices from the context of which products to produce and not produce.
f. Using the results from e), (i.e. reduced cost and shadow prices) make one change to the base model's **objective function** that will change the production plan. Rerun and discuss the new results.
g. Using the results from e), (i.e. reduced cost and shadow prices) make one change to the base model's **resource usage values** that will change the production plan. Rerun and discuss the new results.
h. Using the results from e), (i.e. reduced cost and shadow prices) make one change to the base model's **available resource values** that will change the production plan. Rerun and discuss the new results.
i. Combine the results of the base case e), as well as the variations f) through h) into a single table and discuss the results.

Exercise 4.2 (Revisiting Transportation). Using sensitivity analysis, revisit the transportation exercise from chapter 3.

a. If one more unit of supply was available, where would it be prioritized and why?
b. If demand could be increased by one unit, would it affect the result and at which destination node it be preferred and why?

Exercise 4.3 (Identifying Hidden Costs). In a TcDonald's restaurant, given below is the data for staff time in minutes to perform different steps in order to make Burgers, Coffee, Ice cream and Fries. These steps include: Order receiving, Processing, Preparing, Packaging, and Delivery.

Time (Mins.)	Burger	Coffee	Ice cream	Fries	Available Minutes
Profit	\$2	\$2	\$3	\$1	
Order receiving	2	2	5	1	1200
Processing	2	1	1	1	1500
Preparing	12	6	2	10	3000
Packaging	3	2	4	2	1800
Delivery	1	3	1	1	1200

a. Use R Markdown to create your own description of the model.
b. Extend the R Markdown to show your LP Model. Be sure to define models.
c. Solve the model in R (To find maximum profit).
d. Interpret and discuss the model in R Markdown.

e. Examine and reflect upon the reduced costs and shadow prices. Discuss which step's available hours need to increases/decrease to make the most profit.

5

Data Envelopment Analysis

5.1 Introduction

Data envelopment analysis or DEA is a powerful tool for conducting studies of efficiency and has been used in thousands of publications since its inception in the 1970s.[1]

While tools exist for conducting the evaluations, it is important to understand how the tools work. Many DEA studies have been conducted and published by authors with only a superficial understanding of the technique. This is equivalent to having a house built by carpenters that only (barely?) understand how to use a hammer. The purpose of this document is to show how DEA works, what it means, and how to use R for getting started with DEA. In order to keep things simple, this first step only looks at the input-oriented envelopment model with constant returns to scale. We introduce other models at the end of this chapter.

This chapter walks through how DEA works and then shows how to implement the model in R using two very different approaches. Over the years, I have built DEA models in many languages and platforms: Pascal, LINDO, LINGO, Excel Macros, Excel VBA, GAMS, AMPL, XPress-MOSEL, and GLPK among others.

In this chapter, we will use a few functions from an R package, `TRA`, that I have created. This package is for doing a range of functions related to Technology, Research, and Analytics. The `TRA` package is not posted on CRAN but is instead available from github. You can download a current version from github. The result is that you will need to install it using the `devtools` package.

[1]This chapter is drawn from an introduction chapter in the book, *Data Envelopment Analysis Using R* by the same author. More details on DEA are available from that book. This book is also available via github at `https://github.com/prof-anderson`

$$github.com/prof-anderson$$

DOI: 10.1201/9781003051251-5

```
library (devtools)
devtools::install_github("prof-anderson/TRA")
library (TRA)
```

5.2 Creating the Data

Let's start by defining data. DEA applications can have multiple inputs and outputs. The input(s) are typically resources that are consumed in the production of output(s). Inputs are referred to as "bads" in that higher levels at the same level of output is considered worse. Similarly, holding everything else constant, an increase in any single output is laudable. Examples of inputs might include capital, labor, or number of machines.

In contrast, outputs are the "good" items that are being produced by the actions of the producers. Examples of outputs might include automobiles produced, customers served, or graduating students.

Let's start by creating a simple application – assume that you are a regional manager of a grocery chain. You have four stores and would like to assess the performance of the store managers using number of employees, x, as an input and the weekly sales, y, as an output.

As the manager responsible for the stores in the region, there are many questions that you may want answered:

- Which of these stores are the best?
- Which stores are laggards?
- For those that are underperforming, are there objective performance targets they should be able to reach?
- Which stores might be employing best practices that could be adopted by other stores?
- Among the stores that are lagging in performance, are there particular stores they should focus attention on for learning best practices?

We'll start by drawing a visual representation of the input-output model. The function `DrawIOdiagram` is from the `TRA` package and allows for creating a nicely formatted input-output diagram.

```
library (TRA) # Note, may need to install directly from github
              # remotes::install_github("prof-anderson/TRA")

Figure <- DrawIOdiagram (c("Employees\n(FTE)" ),
                         c("Sales\n($/week)"),
                         '"\nStore\n "')
tmp<-capture.output(rsvg_png(charToRaw(export_svg(Figure)),
                             'images/DEA_IO_stores.PNG'))
knitr::include_graphics("images/DEA_IO_stores.PNG")
```

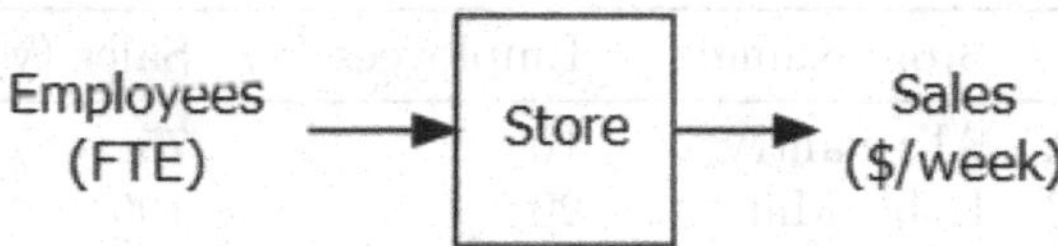

FIGURE 5.1 A Simple DEA Model for Store Management.

The figure for drawing input-output diagrams is worth a little discussion and explanation.

This follows three steps to work for both HTML & PDF:

1. Generate diagram object and save to Figure object
2. Export Figure as image to images directory
3. Display image

Note that it assumes that there is a subdirectory for images – if there isn't a subdirectory for images, you can create one or edit the path.

Furthermore, note that the text passed to the function such as `"Sales\n($/week)"` allows for a line return character such as `"\n"`

Let's get started now with R.

```
x <- matrix(c(10,20,30,50),ncol=1,
            dimnames=list(LETTERS[1:4],"x"))
y <- matrix(c(75,100,300,400),ncol=1,
            dimnames=list(LETTERS[1:4],"y"))

storenames<- c("Al\'s Pantry", "Bob\'s Mill",
               "Trader Carrie\'s", "Dilbertson\'s")
```

```
temp<-cbind(storenames,x,y)
colnames(temp)<-c("Store Name", "Employees (x)", "Sales (y)")
```

```
kbl (temp, booktabs=T, escape=F,
    caption="First Dataset for DEA") |>
  kable_styling(latex_options = "hold_position")
```

TABLE 5.1 First Dataset for DEA

	Store Name	Employees (x)	Sales (y)
A	Al's Pantry	10	75
B	Bob's Mill	20	100
C	Trader Carrie's	30	300
D	Dilbertson's	50	400

The above commands create matrices that hold the data and have named rows and columns to match. The `<-` symbol is a key function in R and means to assign what is in the right to the object on the left.

For benchmarking, we want to know which ones are doing the best job.

Can you tell which stores represent the best tradeoff between inputs and outputs? None of the stores are strictly dominated by any of the other stores. Dominance would be producing more outputs using less input so let's move on to looking at it graphically.

5.3 Graphical Analysis

Let's start by doing a simple plot of the data. For now, I'm going to make use of a function in Peter Bogetoft and Lars Otto's `Benchmarking` package (Bogetoft and Otto, 2013) which provides a very handy and nicely formatted two-dimensional plot in the format often used for showing production. Many of the functions in the package are also described in their book.

```
library(Benchmarking, quietly=TRUE)
dea.plot(x, y, RTS="crs", ORIENTATION="in-out",
```

```
        txt=LETTERS[1:length(x)],
        add=FALSE, wx=NULL, wy=NULL, TRANSPOSE=FALSE,
        xlab="Employees (x)", ylab="Sales (y)",
        fex=1, GRID=TRUE, RANGE=FALSE, param=NULL)
```

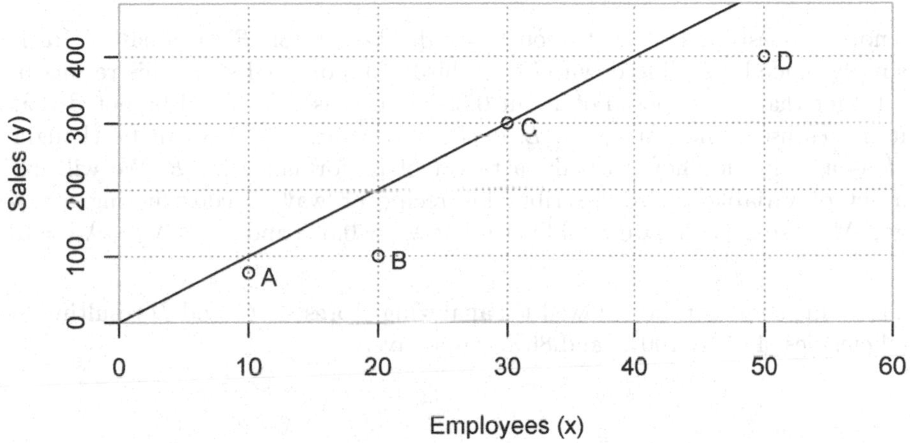

FIGURE 5.2 Store Benchmarking Example.

Note that the function call `dea.plot` from `Benchmarking` uses options from the standard `r` function `plot`. This includes items such as axis labeling `xlab` and `ylab` along with the figure title `main`. It also provides DEA specific options such as `RTS`, `ORIENTATION`, `add`, `wx`, `wy`, and more. The various options for `plot` and `dea.plot` can be found in their respective help pages.

This chart clearly shows that store C (Trader Carrie's) has the best ratio of sales (output, y) to employees (input, x). The diagonal line represents an efficiency frontier of best practices that could be achieved by scaling up or down store C. As the input is scaled up or down, it is assumed that the output of store C would be scaled up or down by the same value. We will revisit this assumption in a later section but for now, think of this as saying that store C cannot enjoy economies of scale by getting larger or suffer from diseconomies of scale by getting smaller, so it is referred to as constant returns to scale or CRS.

Furthermore, we can graphically examine the technical efficiency of each of the other stores. I'm going to start with store B since it is a little easier to visualize. For now, let's think of the question, how much more or less input would C require to produce as much output as B.

To determine this efficiency score, simply draw a horizontal line from B to the efficiency frontier on the left. This point can be thought of as a target for B to be efficient. This point has the same output as B but uses only half as much input. The efficiency score can be calculated as the ratio of the distance from the vertical axis to the target divided by the distance from the vertical axis to B. This distance is simply 10/20 or 50%.

Another question is how to construct the target for B's evaluation. It is simply made by scaling down C to a third of its original size. This results in a target that is composed of about 0.333 of C. Also, it should be noted that it makes use of no part of A, B, or D. This value of 0.333 can be thought of as a recipe for how to make a target of performance for B. We will use a set of variables, λ to describe this recipe or way of constructing a target. More specifically, we would note that $\lambda_C = 0.333$ and $\lambda_A = \lambda_B = \lambda_D = 0$.

The same steps can be followed for analyzing stores A, C, and D resulting in efficiencies of 75%, 100%, and 80%, respectively.

5.4 The Linear Programs for DEA

5.4.1 An Explicit Linear Program for DEA

The graphical approach is intuitive but accuracy is limited to that of drawing tools. More importantly, it does not scale to more complex models with multiple inputs and outputs. Let's frame this topic mathematically so that we can proceed systematically. Furthermore, benchmarking of stores is only one application area of literally thousands so rather than limiting ourselves to stores, we will refer to them as *decision making units* (DMUs) or units for short.

A key way to begin the mathematical development of the envelopment model is to ask, can you find a combination of units that produces a target with at least as much output using less input? The blend of other units is described by a vector λ. Another way to denote this is λ_j, which is the specific amount of a unit j used in setting the target for performance for the unit being studied, k. In this case, k=2 since we are analyzing the second store, B. Furthermore, we can state that we want to find how much less input our target for B needs compared with what B actually used. This value for input usage is a unitless variable, θ where $\theta = 1$ indicates that a reduction cannot occur for each input while maintaining B's level of output. Similarly, a value of $\theta = 0.9$ indicates a 10% reduction in input usage could be obtained.

We could express the evaluation of store B as a linear program as follows.

$$
\begin{aligned}
&\text{Min } \theta && \text{[Efficiency]}\\
&\text{S.t.: } 10\lambda_A + 20\lambda_B + 30\lambda_C + 50\lambda_D \leq 20\theta && \text{[Input]}\\
&\quad 75\lambda_A + 100\lambda_B + 300\lambda_C + 400\lambda_D \geq 100 && \text{[Output]}\\
&\quad \theta,\ \lambda_A,\ \lambda_B,\ \lambda_D,\ \lambda_D \geq 0
\end{aligned}
$$

Implementing this formulation in `ompr` or another system is straightforward and similar to the production planning linear programs examined in earlier chapters.

To solve for other stores, we would simply change the values on the right-hand side of the constraints and solve again. This means that we solve as many linear programs as there are units to evaluate. We are in effect iterating for *k=1,...,Number of Units* or with the four store case, *k=1,...,4*.

5.4.2 A Generalized Linear Program for DEA

This can be easily expanded to the multiple input and multiple output case by defining $x_{i,j}$ to be the amount of the i'th input used by unit j and $y_{r,j}$ to be the amount of the r'th output produced by unit j. For simplicity, this example will focus on the one input and one output case rather than the m input and s output case but the R code explicitly allows for $m, s > 1$. To make the code more readable, I will use a slightly different convention N^X or `NX` instead of m to refer to the number of inputs (x's) and N^Y or `NY` to be the number of outputs (y's) instead of s. Also, the normal mathematical convention is to use n to denote the number of Decision Making Units (DMUs) so I will use N^D or `ND` to indicate that in the R code.

Rather than jumping straight to the linear program, let's take a step back and focus on the construction of the target of performance.

The core idea of the envelopment model of a DMU k can be thought of as to find a target constructed of a mix of the DMU's described by a vector λ that uses no more of any input to achieve the same or more of every output as DMU k. The amount of the *i'th* input used by the target is then $\sum_{j=1}^{N^D} x_{i,j}\lambda_j$.

By the same token, the amount of the *r'th* output produced by the target is $\sum_{j=1}^{N^D} y_{r,j}\lambda_j$.

This gives us two sets of constraints along with a restriction of non-negativity. These are shown in the following relationships that must all be satisfied simultaneously.

$$\sum_{j=1}^{N^D} x_{i,j}\lambda_j \leq x_{i,k} \; \forall \; i \quad \text{[Use no more input than k]}$$

$$\sum_{j=1}^{N^D} y_{r,j}\lambda_j \geq y_{r,k} \; \forall \; r \quad \text{[At least as much output]}$$

$$\lambda_j \geq 0 \; \forall \; j$$

This is not yet a linear program because it is missing an objective function. It defines what is an acceptable target of performance that is at least as good as DMU k but does not try to find a *best* target.

The two most common approaches to finding the best target are the input-oriented and output-oriented models. In the output-oriented model, the first (input) constraint is satisfied while trying to *exceed* the second constraint (output) by as much possible. This focus on increasing the output is then called an *output orientation.*

In this chapter, we will focus on satisfying the second constraint while trying to improve upon the first by as much as possible. In other words, we will satisfy the second (output) constraint but try to form a target that uses as little input as possible. The focus on reducing inputs gives it the name of the input-oriented model.

In fact, we will go one step further and say that we want to find the maximum possible input reduction in k's input or conversely, the minimum amount of the input that could be used by the target while still producing the same or more output. We do this by adding a new variable, θ, which is the radial reduction in the amount of DMU k's input. We want to find how low we can drive this by *minimizing* θ. Let's define the proportion of the studied unit's input needed by the target as θ. A value of $\theta = 1$ then means no input reduction can be found in order to produce that unit's level of output.

This gives us the following linear program.

$$\begin{aligned}
&\min \theta \\
&\text{s.t.: } \sum_{j=1}^{N^D} x_{i,j}\lambda_j \leq \theta x_{i,k} \ \forall \ i \\
&\sum_{j=1}^{N^D} y_{r,j}\lambda_j \geq y_{r,k} \ \forall \ r \\
&\lambda_j \geq 0 \ \forall \ j
\end{aligned}$$

Expressing the target on the left and the actual unit's value and radial reduction on the right is conceptually straightforward to understand. Some optimization software requires collecting all the variables on the left and putting constants on the right-hand side of the inequalities. This is easily done and is shown in the following linear program for completeness but a benefit of `ompr` is that we can specify the model in the original format.

$$\begin{aligned}
&\min \theta \\
&\text{s.t.: } \sum_{j=1}^{N^D} x_{i,j}\lambda_j - \theta x_{i,k} \leq 0 \ \forall \ i \\
&\sum_{j=1}^{N^D} y_{r,j}\lambda_j \geq y_{r,k} \ \forall \ r \\
&\lambda_j \geq 0 \ \forall \ j
\end{aligned}$$

5.5 Creating the LP – The Algebraic Approach

There are two fundamentally different approaches to setting up linear programs for solving. The first approach is to define data structures to pass vectors for the objective function coefficients and constraint right-hand sides along with a matrix of data describing the constraints. This requires careful setting up of the linear programs and is a big cognitive step away from the mathematical representation. Another approach is to use algebraic modeling languages. Standalone algebraic optimization modeling languages include LINGO, AMPL, GAMS, GMPL, and others.

Until recently, R did not have the ability to do algebraic modeling optimization but a few new packages have provided support for this. In particular, `ompr`, provides an algebraic perspective that matches closely to the summation representation of a linear program shown earlier. Don't worry, if you want to see the data structure format approach, that is covered in the *DEA Using R* book.

Let's define some data structures for holding our data and results.

```
ND <- nrow(x); NX <- ncol(x); NY <- ncol(y);
                # Define data size
xdata<-x[1:ND,]
dim(xdata)<-c(ND,NX)
ydata<-y[1:ND,]
dim(ydata)<-c(ND,NY)
                # Now we will create lists of names
DMUnames <- list(c(LETTERS[1:ND]))
   # DMU names: A, B, ...
Xnames<- lapply(list(rep("X",NX)),paste0,1:NX)
   # Input names: x1, ...
Ynames<- lapply(list(rep("Y",NY)),paste0,1:NY)
   # Output names: y1, ...
Vnames<- lapply(list(rep("v",NX)),paste0,1:NX)
   # Input weight names: v1, ...
Unames<- lapply(list(rep("u",NY)),paste0,1:NY)
   # Output weight names: u1, ...
SXnames<- lapply(list(rep("sx",NX)),paste0,1:NX)
   # Input slack names: sx1, ...
SYnames<- lapply(list(rep("sy",NY)),paste0,1:NY)
   # Output slack names: sy1, ...
Lambdanames<- lapply(list(rep("L_",ND)),
                     paste0,LETTERS[1:ND])
results.efficiency <- matrix(rep(-1.0, ND),
                             nrow=ND, ncol=1)
dimnames(results.efficiency)<-c(DMUnames,"CCR-IO")
   # Attach names

results.lambda <- matrix(rep(-1.0, ND^2),
                         nrow=ND,ncol=ND)
dimnames(results.lambda)<-c(DMUnames,Lambdanames)
results.xslack     <- matrix(rep(-1.0, ND*NX),
                             nrow=ND,ncol=NX)
dimnames(results.xslack)<-c(DMUnames,SXnames)
results.yslack     <- matrix(rep(-1.0, ND*NY),
                             nrow=ND,ncol=NY)
dimnames(results.yslack)<-c(DMUnames,SYnames)
```

We're going to use our data from earlier but first we will load a collection of libraries to be used later. The omprpackage is for optimization and serves as a general human readable format of optimization models that can then interface with a variety of solver engines. The ROI.plugin.glpk package is for the specific solver engine, glpk, that we used. Other LP solving engines are available and can be used instead. We demonstrate other solvers in chapter 7.

```
library(dplyr, quietly=TRUE)                # For data organizing
library(ROI, quietly=TRUE)                  # R Optimization Interface
library(ROI.plugin.glpk, quietly=TRUE) # Connection to glpk as solver
library(ompr, quietly=TRUE)                 # Optimization Modeling using R
library(ompr.roi, quietly=TRUE)             # Connective tissue
```

Now that we have loaded all of the packages that we use as building blocks, we can start constructing the model.

We are going to start by building a model for just one DMU, in this case, the second DMU (B).

```
k<-2     # DMU to analyze.
         # Let's start with just one DMU, B, for now.
result <- MIPModel()                                              |>
  add_variable(vlambda[j], j = 1:ND, type = "continuous",
               lb = 0)                                            |>
  add_variable(vtheta, type = "continuous")                       |>
  set_objective(vtheta, "min")                                    |>
  add_constraint(sum_expr(vlambda[j] * xdata[j,1], j = 1:ND)
                 <= vtheta * xdata[k,1])                          |>
  add_constraint(sum_expr(vlambda[j] * ydata[j,1], j = 1:ND)
                 >= ydata[k,1])                                   |>
  solve_model(with_ROI(solver = "glpk"))
omprtheta <-  get_solution(result, vtheta)
omprlambda <-  get_solution(result, vlambda[j])
ND <- 4 # Four Decision Making Units or DMUs
NX <- 1 # One input
NY <- 1 # One output
   # Only doing analysis for one unit at a time to start
results.efficiency <- matrix(rep(-1.0, 1), nrow=1, ncol=1)
results.lambda     <- matrix(rep(-1.0, ND), nrow=1,ncol=ND)
results.efficiency <- t(omprtheta)
colnames(results.efficiency) <- c("CCR-IO")
```

```
results.lambda <- t(omprlambda[3])
   # Takes third column from results and transposes results
   #  to be structured correctly for later viewing
results_B <- cbind (results.efficiency, results.lambda)
rownames(results_B) <- c("Optimal results for DMU B")

kbl(results_B, digits=4,booktabs=T, escape=F,
    col.names=c("$\\theta^{CRS}$", "$\\lambda_A$",
                "$\\lambda_B$", "$\\lambda_C$", "$\\lambda_D$"),
    caption="Input-Oriented Envelopment Analysis for DMU B") |>
  kable_styling(latex_options = "hold_position")
```

TABLE 5.2 Input-Oriented Envelopment Analysis for DMU B

	θ^{CRS}	λ_A	λ_B	λ_C	λ_D
Optimal results for DMU B	0.5	0	0	0.3333	0

```
# Note: Creating LaTeX column names is covered in appendix D.
```

The above table follows a few convenient conventions. First, the θ, gets a superscript to indicate that it for the CRS or constant returns to scale model. This is sometimes labeled as CCR after Charnes, Cooper, and Rhodes (Rhodes, 1978). Later we will cover other models including the variable returns to scale model, sometimes labeled BCC after Banker, Charnes, and Cooper (Banker et al., 1984).

Another convention is to embed the *orientation* in the table. This an input-oriented model where the primary focus is on achieving efficiency through input reduction. Output-orientation is also very common where the primary goal is for a unit to increase output with the requirement of not using any more input.

The results in the table indicate that DMU B has an efficiency score of 50%. The target of performance is made of DMU C scaled down by a factor of 0.33. These results match the graphical results from earlier. This 50% efficiency score means that if B were using best practices as demonstrated by the other units, it would only need 50% of the inputs to produce the same output.

Another item to note is that we are using some of the flexibility of `kable` in our table. In particular, we are able to use LaTeX in our in column headings as long as we set `escape=F`. Note from the code chunk that the actual use requires an extra slash.

Let's now extend it to handle multiple inputs, NX, and outputs, NY. Of course this doesn't have any impact on our results just yet since we are still only using a single input and output but we now have the structure to accommodate the more general case of DEA. To provide a little variety, we'll change it to the first DMU, A, to give a little more variety.

```
k<-1    # Analyze first unit, DMU A.
result <- MIPModel()                                                    |>
  add_variable(vlambda[j], j = 1:ND, type = "continuous",
               lb = 0)                                                  |>
  add_variable(vtheta, type = "continuous")                             |>
  set_objective(vtheta, "min")                                          |>
  add_constraint(sum_expr(vlambda[j] * xdata[j,i], j = 1:ND)
                 <= vtheta * xdata[k,i], i = 1:NX)                      |>
  add_constraint(sum_expr(vlambda[j] * ydata[j,r], j = 1:ND)
                 >= ydata[k,r], r = 1:NY)                               |>
  solve_model(with_ROI(solver = "glpk"))

omprtheta <-  get_solution(result, vtheta)
omprlambda <-  get_solution(result, vlambda[j])
results.efficiency <- t(omprtheta)
colnames(results.efficiency) <- c("CCR-IO")
results.lambda <- t(omprlambda[3])

# Note the use of LaTeX formatting in column names.

results_A <- cbind (results.efficiency, results.lambda)
rownames(results_A) <- c("Optimal results for DMU A")

kbl(results_A, booktabs=T, escape=F,
    col.names= c("$\\theta^{CRS}$", "$\\lambda_A$",
                 "$\\lambda_B$", "$\\lambda_C$", "$\\lambda_D$"),
    caption="Input-Oriented Envelopment Analysis for DMU A")  |>
  kable_styling(latex_options = "hold_position")
```

TABLE 5.3 Input-Oriented Envelopment Analysis for DMU A

	θ^{CRS}	λ_A	λ_B	λ_C	λ_D
Optimal results for DMU A	0.75	0	0	0.25	0

Again, the results match what would be expected the graphical analysis.

Now we should extend this to handle all four of the decision making units. A key new function that we use here is the `for` command to loop the previous code that we had used for analyzing A and B separately. Notably, we assign the results to matrices at the end of each loop.

```
results.efficiency <- matrix(rep(-1.0, 1), nrow=ND, ncol=1)
results.lambda     <- matrix(rep(-1.0, ND), nrow=ND,ncol=ND)
for (k in 1:ND) {
  result <- MIPModel()                                    |>
    add_variable(vlambda[j], j=1:ND, type = "continuous",
                 lb = 0)                                  |>
    add_variable(vtheta, type = "continuous")             |>
    set_objective(vtheta, "min")                          |>
    add_constraint(sum_expr(vlambda[j]*xdata[j,i], j=1:ND)
                   <= vtheta * xdata[k,i], i = 1:NX)      |>
    add_constraint(sum_expr(vlambda[j] * ydata[j,r], j=1:ND)
                   >= ydata[k,r], r = 1:NY)               |>
    solve_model(with_ROI(solver = "glpk"))

  print(c("DMU=",k,solver_status(result)))
  results.efficiency[k] <-  get_solution(result, vtheta)
  results.lambda[k,] <- t(as.matrix(as.numeric(
    get_solution(result, vlambda[j])[,3] )))
}
```

```
## [1] "DMU="    "1"        "optimal"
## [1] "DMU="    "2"        "optimal"
## [1] "DMU="    "3"        "optimal"
## [1] "DMU="    "4"        "optimal"
```

Success! This indicates each of the four linear programs was solved to optimality. By itself, it doesn't help much though. We need to now display each column of results. Lambda, λ, reflects the way that a best target is made for that unit.

```
Lambdanames <- list("$\\lambda_A$", "$\\lambda_B$",
                    "$\\lambda_C$", "$\\lambda_D$")
DMUnames <- list("A", "B", "C", "D")
dimnames(results.efficiency)<-list(DMUnames,"CCR-IO")
eff.crs <- results.efficiency
dimnames(results.lambda)<-list(DMUnames,Lambdanames)
kbl (cbind(results.efficiency, results.lambda),
```

```
  booktabs=T, escape=F, digits=4, caption=
    "Input-Oriented Efficiency Results") |>
kable_styling(latex_options = c("hold_position"))
```

TABLE 5.4 Input-Oriented Efficiency Results

	CCR-IO	λ_A	λ_B	λ_C	λ_D
A	0.75	0	0	0.2500	0
B	0.50	0	0	0.3333	0
C	1.00	0	0	1.0000	0
D	0.80	0	0	1.3333	0

TABLE 5.5 Results with Inefficient DMU Columns Removed

	CCR-IO	λ_C
A	0.75	0.2500
B	0.50	0.3333
C	1.00	1.0000
D	0.80	1.3333

The results match those observed graphically but let's discuss them. These results indicate that only DMU C is efficient. Rescaled versions of C could produce the same level of output of A, B, or D, while using just 25%, 50%, and 20% less input, respectively. The targets of performance for A and B are constructed by scaling down unit C to a much smaller size, as shown by the values of λ. In contrast, D's performance is surpassed by a 33% larger version of C.

It is an important step of any DEA study to carefully examine the results. In this case, it might be argued that for certain applications, C's business practices do not readily scale, and therefore, could not be assumed to operate at a much smaller or larger size. Recall our original application of stores. It might be that practices employed by store C, Trader Carrie's, do not easily scale to larger or smaller operations. Al's Pantry (store A) is a much smaller operation and just trying to keep the store in operation with ten employees may be a major challenge. Perhaps all of the stores offer 24 hour a day operation. During the slow time after midnight, Trader Carrie's could have just one employee working. Meanwhile Al's Pantry could not stay in business with one third of an employee working at 3 AM!

Similarly, Trader Carrie's may not scale to larger operations. In a mid-sized store, employees may feel more comfortable with all parts of the store but in a larger store, people work in separate departments such as the produce department and the meat department. This separation by departments may yield benefits or disbenefits.

Up until this point, we have assumed constant returns to scale but there are other versions of returns to scale to better fit the needs of each application. Next, let's incorporate modeling returns to scale.

5.6 Returns to Scale

Let's also add a constraint that will accommodate returns to scale. All it needs to do is constrain the sum of the λ variables equal to 1 enforces variables returns to scale or VRS. or redundant under CRS.

$$\begin{aligned}
\text{minimize } & \theta \\
\text{subject to } & \sum_{j=1}^{n} \lambda_j = 1 \\
& \sum_{j=1}^{n} x_{i,j}\lambda_j - \theta x_{i,k} \leq 0 \ \forall \ i \\
& \sum_{j=1}^{n} y_{r,j}\lambda_j \geq y_{r,k} \ \forall \ r \\
& \lambda_j \geq 0 \ \forall j
\end{aligned}$$

To incorporate this, we can add another constraint to our previous model and solve it. Let's define a parameter, "RTS" to describe which returns to scale assumption we are using and only add the VRS constraint when RTS="VRS".

The other very common returns to scale option is constant returns to scale or CRS, which is what we have used up to this point. For CRS, you can delete the VRS constraint, but it may be helpful in some implementations to maintain a consistent model size for reading out sensitivity information. To maintain the number of rows (constraints) we can make it a redundant constraint by constraining the sum of λ to be greater than or equal to zero, $\sum_{j=1}^{n} \lambda_j \geq 0$ Since λ's are by definition non-negative, the sum of λ's is also non-negative, and therefore, the constraint is superfluous or redundant.

```
RTS<-"VRS"
for (k in 1:ND) {

  result <- MIPModel()                                          |>
    add_variable(vlambda[j], j = 1:ND, type = "continuous",
                 lb = 0)                                        |>
    add_variable(vtheta, type = "continuous")                   |>
    set_objective(vtheta, "min")                                |>
    add_constraint(sum_expr(vlambda[j] * xdata[j,i], j = 1:ND)
                   <= vtheta * xdata[k,i], i = 1:NX)            |>
    add_constraint(sum_expr(vlambda[j] * ydata[j,r], j = 1:ND)
                   >= ydata[k,r], r = 1:NY)
    if (RTS=="VRS") {result <- add_constraint(result,
                       sum_expr(vlambda[j], j = 1:ND)
                       == 1)
                     } #Returns to Scale
result <- solve_model(result, with_ROI (solver = "glpk"))

    results.efficiency[k] <-  get_solution(result, vtheta)
    results.lambda[k,] <- t(as.matrix(as.numeric(
              get_solution(result, vlambda[j])[,3] )))
} # Repeat for each unit, k
dimnames(results.efficiency)<-list(DMUnames,"$\\theta^{VRS}$")
eff.vrs <- results.efficiency
dimnames(results.lambda)<-list(DMUnames,Lambdanames)
kbl (cbind(results.efficiency, results.lambda),
     booktabs=T, escape=F, digits=4, caption=
       "Input-Oriented VRS Envelopment Results")                |>
  kable_styling(latex_options = "hold_position")
```

TABLE 5.6 Input-Oriented VRS Envelopment Results

	θ^{VRS}	λ_A	λ_B	λ_C	λ_D
A	1.0000	1.0000	0	0.0000	0
B	0.6111	0.8889	0	0.1111	0
C	1.0000	0.0000	0	1.0000	0
D	1.0000	0.0000	0	0.0000	1

Notice that the efficiencies have generally increased or stayed the same. Whereas earlier three out of four DMUs were inefficient, now three out of four are efficient. One way of thinking of returns to scale is whether doubling the inputs should be expected to result in doubling the outputs that should be achieved. Another way to think of it is whether it is fair to think of scaling up or down an efficient significantly to set a performance target for a much bigger or smaller unit. For example, would it be *fair* to compare a small convenience store such as Al's Pantry or to use a common realworld example, 7-11, to a CostCo store scaled down by a factor of a 100? Much more could be said about returns to scale.

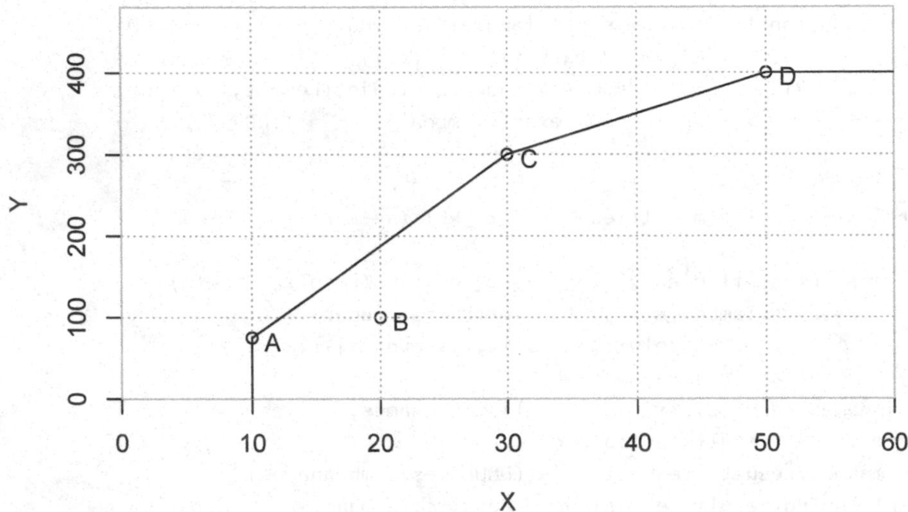

FIGURE 5.3 BCC (VRS) Efficiency Frontier.

The Constant Returns to Scale, CRS, model is often referred to in the DEA literature by CCR after Charnes, Cooper, and Rhodes (Charnes et al., 1978). The Variable Returns to Scale, VRS, model is also referred to as the BCC model after Banker, Charnes, and Cooper. In addition to the CRS and VRS models, two other common approaches are Increasing Returns to Scale (IRS) and Decreasing Returns to Scale (DRS). Technically, IRS is sometimes more precisely referred to as non-decreasing returns to scale. Similarly, DRS corresponds to non-increasing returns to scale. A less commonly used approach is Free Disposal Hull (FDH). In the case of FDH, DMUs are only compared against other individual units. Since it makes use of binary variables and is less commonly used in DEA, we will not implement it in the chapter.

TABLE 5.7 Returns to Scale Envelopment Constraints

Returns to Scale	Envelopment Constraint
CRS	No constraint needed
VRS	$\sum_{j=1}^{N^D} \lambda_j = 1$
IRS/NDRS	$\sum_{j=1}^{N^D} \lambda_j \geq 1$
DRS/NIRS	$\sum_{j=1}^{N^D} \lambda_j \leq 1$
FDH	$\sum_{j=1}^{N^D} \lambda_j = 1, \lambda_j \in \{0,1\} \; \forall \; j$

Now, we will generalize this by allowing a parameter to set the returns to scale.

```
RTS<-"DRS"
for (k in 1:ND) {

  result <- MIPModel()                                                 |>
    add_variable(vlambda[j], j = 1:ND, type = "continuous",
                 lb = 0)                                               |>
    add_variable(vtheta, type = "continuous")                          |>
    set_objective(vtheta, "min") |>
    add_constraint(sum_expr(vlambda[j] * xdata[j,i], j = 1:ND)
                   <= vtheta * xdata[k,i], i = 1:NX)                   |>
    add_constraint(sum_expr(vlambda[j] * ydata[j,r], j = 1:ND)
                   >= ydata[k,r], r = 1:NY)
    if (RTS=="VRS") {result <-
      add_constraint(result, sum_expr(vlambda[j], j = 1:ND) == 1) }
    if (RTS=="IRS") {result <-
      add_constraint(result, sum_expr(vlambda[j], j = 1:ND) >= 1) }
    if (RTS=="DRS") {result <-
      add_constraint(result, sum_expr(vlambda[j], j = 1:ND) <= 1) }
result <- solve_model(result, with_ROI(solver = "glpk"))

    results.efficiency[k] <-  get_solution(result, vtheta)
    results.lambda[k,] <- t(as.matrix(as.numeric(
              get_solution(result, vlambda[j])[,3] )))
}
dimnames(results.efficiency)<-list(DMUnames,"DRS-IO")
eff.drs <- results.efficiency
dimnames(results.lambda)<-list(DMUnames,Lambdanames)
kbl (cbind(results.efficiency, results.lambda),
     booktabs=T,  escape=F, digits=4,
```

```
    caption="Input-Oriented Model with
      Decreasing Returns to Scale") |>
  kable_styling(latex_options = "hold_position")
```

TABLE 5.8 Input-Oriented Model with Decreasing Returns to Scale

	DRS-IO	λ_A	λ_B	λ_C	λ_D
A	0.75	0	0	0.2500	0
B	0.50	0	0	0.3333	0
C	1.00	0	0	1.0000	0
D	1.00	0	0	0.0000	1

Simply changing from `RTS<-"DRS"` to `RTS<-"IRS"` in the first line allows to now evaluate the Increasing Returns to Scale case.

TABLE 5.9 Input-Oriented Increasing Returns to Scale Model Results

	IRS-IO	λ_A	λ_B	λ_C	λ_D
A	1.0000	1.0000	0	0.0000	0
B	0.6111	0.8889	0	0.1111	0
C	1.0000	0.0000	0	1.0000	0
D	0.8000	0.0000	0	1.3333	0

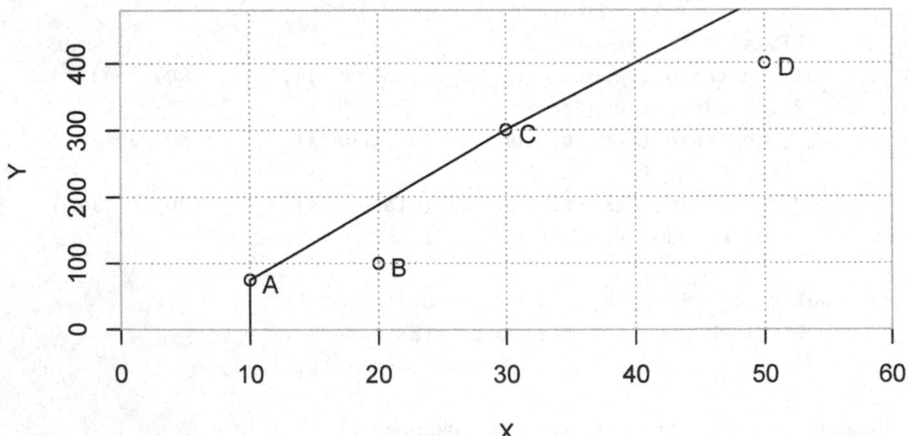

FIGURE 5.4 Increasing (Non-Decreasing) Returns to Scale Efficiency Frontier.

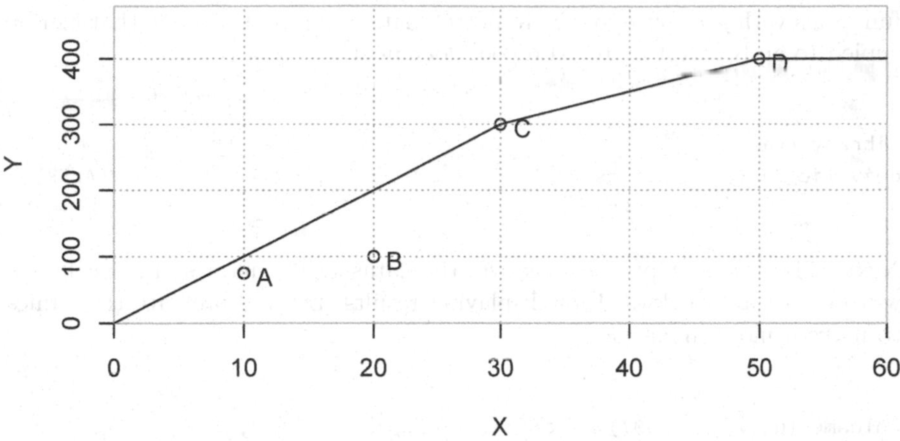

FIGURE 5.5 Decreasing(Non-Increasing) Returns to Scale Efficiency Frontier.

TABLE 5.10 Comparison of Efficiency under Alternate Returns to Scale Assumptions

	θ^{CRS}	θ^{VRS}	θ^{DRS}	θ^{IRS}
A	0.75	1.0000	0.75	1.0000
B	0.50	0.6111	0.50	0.6111
C	1.00	1.0000	1.00	1.0000
D	0.80	1.0000	1.00	0.8000

5.7 Multiple Inputs and Multiple Outputs

Using DEA for two-dimensional examples such as the one-input, one-output model is easy to draw and visualize but overkill and not generally very useful.

Let's move on to a larger application of examining technology transfer success from university research by revisiting (Anderson et al., 2007). This paper used data from the Association of University Technology Managers, AUTM, and their 2004 survey. We will start by loading the data from the paper. The dataset can be accessed as a comma separated file (CSV) format from our github repository. Note that in RStudio, many datasets can be easily loaded using the `Import Dataset`command from under the File pulldown command. This provides a nice graphical user interface wrapper around various importing functions with a preview of the imported data along with R code that can be copied to an R script or RMarkdown document.

```
library(readr)
univ_lic_2007 <- read_csv("univ_lic_2007.csv", show_col_types=FALSE)
```

Next, it is always helpful to look over the dataset. The names of the columns will be long and awkward for displaying results. Let's abbreviate the names to make it more manageable.

```
colnames(univ_lic_2007) <- c("University", "LicInc",
                             "LicOpt", "Startup",
                             "PatFile", "PatIss", "Spend")

utt2007 <- univ_lic_2007 [1:25,] # Use first 25 universities

utt2007 <- utt2007 [,-5] # Drop the fifth column (Patents Filed)

kbl (head(utt2007), booktabs=T,
     caption="Simplified Data for University Technology Transfer") |>
  kable_styling(latex_options = "hold_position")
```

TABLE 5.11 Simplified Data for University Technology Transfer

University	LicInc	LicOpt	Startup	PatIss	Spend
New York University	109.02312	30	4	23	244.4150
U of California System	74.27500	273	5	270	2791.7770
U of Wisconsin at Madison	47.68917	203	2	93	763.8750
Stanford	47.27240	89	9	87	693.5299
U of Minnesota	45.55076	100	3	36	515.0610
U of Florida	37.40228	64	8	53	427.9973

Rather than using the full dataset. We are going to simplify the data a little. First, we limit it to just the first 25 universities instead of the full set of 54. Second, we will drop the measure of patents filed.

Now, let's prepare our data for the upcoming analyses.

```
xdata <- as.matrix(utt2007 [,6])
rownames(xdata)<-as.matrix(utt2007[,1])

ydata <- as.matrix(utt2007 [,2:5])
rownames(ydata)<-as.matrix(utt2007[,1])

Xnames <- colnames(xdata)
Ynames <- colnames(ydata)
DMUnames <-list(as.matrix(utt2007[,1]))

dimnames(xdata)              <- c(DMUnames,Xnames)
colnames(ydata)              <- Ynames

ND <- nrow(xdata)  # Number of DMUs (universities)
NX <- ncol(xdata)  # Number of inputs (just 1 in this case)
NY <- ncol(ydata)  # Number of outputs

res.efficiency <- matrix(rep(-1.0, ND), nrow=ND, ncol=1)
res.lambda      <- matrix(rep(-1.0, ND^2), nrow=ND,ncol=ND)
dimnames(res.efficiency) <- c(DMUnames,"CCR-IO")
dimnames(res.lambda)     <- c(DMUnames,DMUnames)

#Define printable names to be used as appropriate
ynames_printable<-c("Licensing Income\n($M)",
                    "Licenses and \n Options Executed",
                    "Startup Companies",
                    "US Patents Issued")
xnames_printable<-c("Total Research\n Spending ($M)")
DMUnames_printable <- as.matrix(utt2007[,1])
```

As usual, let's preview the data.

```
kbl (head(cbind(xdata,ydata)), booktabs=T,
     caption="Selected University Technology
     Transfer Characteristics.")
```

Now, let's move on to the DEA input-output diagram. We use the total research spending as an input and then five separate outputs for the various dimensions of technology transfer success measured by AUTM. This can be visualized with our input-output diagram.

TABLE 5.12 Selected University Technology Transfer Characteristics

	Spend	LicInc	LicOpt	Startup	PatIss
New York University	244.4150	109.02312	30	4	23
U of California System	2791.7770	74.27500	273	5	270
U of Wisconsin at Madison	763.8750	47.68917	203	2	93
Stanford	693.5299	47.27240	89	9	87
U of Minnesota	515.0610	45.55076	100	3	36
U of Florida	427.9973	37.40228	64	8	53

```
Figure <- DrawIOdiagram (xnames_printable,
                         ynames_printable,
                         '"\n\nUniversity\nTechnology\nTransfer\n\n"')
tmp<-capture.output(rsvg_png(
  charToRaw(export_svg(Figure)),
  'images/DEA_Univ_IO.PNG'))
knitr::include_graphics(
  "images/DEA_Univ_IO.PNG")
```

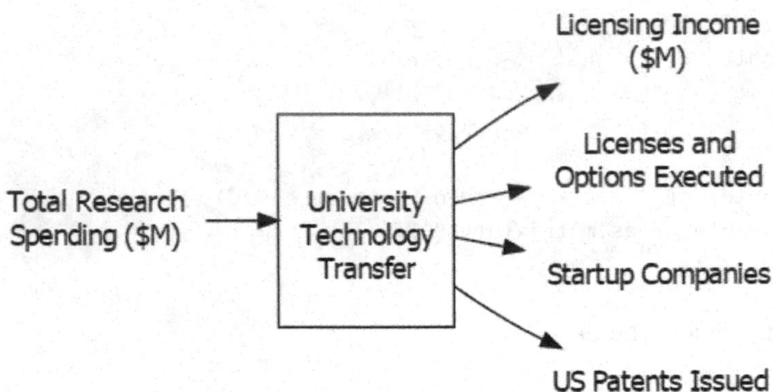


```
RTS<-"CRS"
for (k in 1:ND) {

  modUTT_CCR <- MIPModel() |>
    add_variable(vlambda[j], j = 1:ND, type = "continuous",
                 lb = 0) |>
    add_variable(vtheta, type = "continuous") |>
```

```
    set_objective(vtheta, "min") |>
    add_constraint(sum_expr(vlambda[j] * xdata[j,i], j = 1:ND)
                   <= vtheta * xdata[k,i], i = 1:NX,
                   .show_progress_bar=FALSE ) |>
    add_constraint(sum_expr(vlambda[j] * ydata[j,r], j = 1:ND)
                   >= ydata[k,r], r = 1:NY,
                   .show_progress_bar=FALSE )
    if (RTS=="VRS") {result <- add_constraint(result,
                 sum_expr(vlambda[j],j = 1:ND) == 1) }
        #Returns to Scale
  res <- solve_model(modUTT_CCR, with_ROI(solver = "glpk"))

  res.efficiency[k] <- get_solution(res, vtheta)
  res.lambda[k,] <- t(as.matrix(as.numeric(
                    get_solution(res, vlambda[j])[,3] )))
}
```

```
kbl (cbind(res.efficiency, poscol(res.lambda)),
    booktabs=T, digits=4, escape=F,
    col.names = c("$\\theta^{CRS}$", "$\\lambda_{\\text{NYU}}$",
                  "$\\lambda_{\\text{UW-Madison}}$",
                  "$\\lambda_{\\text{Cal Tech}}$"),
    caption="Results from University Technology
    Transfer Example (CRS)") |>
  kable_styling(latex_options =
                c("hold_position", "scale_down"))
```

```
UTT_CCR.Res<-cbind(res.efficiency, res.lambda)
```

Note that the results are different from those in the UTT paper because the model is different and the data is trimmed. This model is an input-oriented model with constant returns to scale. In contrast, the UTT paper uses an output-oriented variable returns to scale model with a weight restriction relationship. Recall that the data is trimmed to the first 25 universities and patents filed is dropped as less important than patents issued.

Every DMU is compared to one or more of the efficient DMUs. In this case there are three efficient DMUs. These are New York University (NYU), University of Wisconsin at Madison (UW-Madison), and California Institute of

TABLE 5.13 Results from University Technology Transfer Example (CRS)

	θ^{CRS}	λ_{NYU}	$\lambda_{\text{UW-Madison}}$	$\lambda_{\text{Cal Tech}}$
New York University	1.0000	1.0000	0.0000	0.0000
U of California System	0.4663	0.1074	1.0661	1.1858
U of Wisconsin at Madison	1.0000	0.0000	1.0000	0.0000
Stanford	0.6996	0.2621	0.2827	0.5276
U of Minnesota	0.8241	0.2169	0.4407	0.0893
U of Florida	0.8788	0.2226	0.1754	0.4828
Michigan State University	0.7925	0.2574	0.1196	0.2665
Wake Forest University	0.5922	0.3134	0.0000	0.0126
U of Colorado System	0.5443	0.2440	0.0401	0.5674
U of Rochester	0.7543	0.2711	0.0000	0.4225
U of Massachusetts	0.5053	0.1751	0.1352	0.0735
MIT	0.7856	0.0000	0.3547	1.3779
U of Washington	0.4440	0.0654	0.2359	0.4476
Emory University	0.4330	0.1592	0.0908	0.0844
Harvard	0.4208	0.0498	0.1846	0.2451
U. of Utah	0.5948	0.0675	0.1129	0.1789
Florida State U.	0.3989	0.1190	0.0000	0.1357
State U. of New York Research Foundation	0.4153	0.0191	0.1382	0.4748
U. of Texas Southwestern Medical Center at Dallas	0.5318	0.0359	0.1273	0.1573
Case Western Reserve U.	0.5127	0.0685	0.0202	0.2633
U. of Iowa Research Foundation	0.6163	0.0000	0.2066	0.0900
U. of Michigan at Ann Arbor	0.6293	0.0000	0.1588	0.9059
California Tech	1.0000	0.0000	0.0000	1.0000
Washington U. in St. Louis	0.7998	0.0000	0.2303	0.4055
U. of Chicago UCTech	0.3717	0.0271	0.1043	0.0893

Technology (Cal Tech). The `poscol` function from the `TRA` package trimmed out all of the unused columns in the matrix of lambdas. This then left us with just the efficiency scores and the three columns with values of λ for each of these efficient universities. Given the length of the names of the names of the universities, it would still exceed text size. The `kableExtra` package has an option of `scale_down` which makes it easy to fit a table in the printable area.

The University of California System, UC-System, receives an efficiency score of 0.4663. It has a target made up of approximately 0.107 of New York University (NYU), 1.066 of the University of Wisconsin at Madison, 1.186 of Cal Tech that produces similar output using only 47% of the input. This means that if the UC-System were operating efficiently as defined in this model, it would be able to produce the same or more outputs as it actually did using only 47% of the input (Research Funding) as it actually used. This target of performance is made up more more than two other full universities and part of a third.

Note that these values are rounded using the `digits=4` from `kable`. The result of this is it masks odd some values of λ that could warrant a more careful look. In particular, there may be very small positive or negative numbers in the table that are shown as 0.0000. These numbers are due to computational issues in the linear programming solvers and should be interpreted as zero. Rounding the results can make the tables easier for interpretation by the user. The pander packages makes it easy to round numbers to varying levels of precision. The small, nearly zero, values that may appear in DEA calculations can cause some computational difficulty. For example, testing for zero lambda values could miss cases where the value is approximately but not exactly zero.

Perhaps the leader of the UC-System would contend that running a ten campus university system with over a quarter million students is very much unlike running the others added together (although NYU is also a large system.) This could suggest a Variable Returns to Scale model which reflects that it operating at a very large and very small scale provide special challenges.

Let's compare the results from the constant and variable returns to scale cases.

```
RTS<-"VRS"
for (k in 1:ND) {

  modUTT_BCC <- MIPModel()                                         |>
    add_variable(vlambda[j], j = 1:ND, type="continuous",
                 lb = 0)                                           |>
    add_variable(vtheta, type = "continuous")                      |>
    set_objective(vtheta, "min")                                   |>
    add_constraint(sum_expr(vlambda[j] * xdata[j,i], j = 1:ND)
                   <= vtheta * xdata[k,i], i = 1:NX,
                   .show_progress_bar=FALSE )                      |>
    add_constraint(sum_expr(vlambda[j] * ydata[j,r], j = 1:ND,
                   .show_progress_bar=FALSE )
                   >= ydata[k,r], r = 1:NY)
    if (RTS=="VRS") {modUTT_BCC <-
      add_constraint(modUTT_BCC,
                 sum_expr(vlambda[j],j = 1:ND) == 1) }
                 #Returns to Scale
res <- solve_model(modUTT_BCC, with_ROI(solver = "glpk"))

    res.efficiency[k] <- get_solution(res, vtheta)
    res.lambda[k,] <- t(as.matrix(as.numeric(
                        get_solution(res, vlambda[j])[,3] )))
}
#UTT_BCC.Res<-cbind(res.efficiency, res.lambda)
UTT_BCC.Res<-res.efficiency
```

TABLE 5.14 Comparison of CRS vs. VRS Efficiency Scores

	θ^{CRS}	θ^{VRS}
New York University	1.000000	1.000000
U of California System	0.466286	1.000000
U of Wisconsin at Madison	1.000000	1.000000
Stanford	0.699625	0.732835
U of Minnesota	0.824059	0.874977
U of Florida	0.878757	0.907647
Michigan State University	0.792484	0.906014
Wake Forest University	0.592206	1.000000
U of Colorado System	0.544260	0.571191
U of Rochester	0.754255	0.850801
U of Massachusetts	0.505347	0.691422
MIT	0.785563	1.000000
U of Washington	0.444039	0.475481
Emory University	0.432986	0.646545
Harvard	0.420787	0.513422
U. of Utah	0.594784	0.827146
Florida State U.	0.398852	0.790456
State U. of New York Research Foundation	0.415255	0.469800
U. of Texas Southwestern Medical Center at Dallas	0.531787	0.763226
Case Western Reserve U.	0.512656	0.772125
U. of Iowa Research Foundation	0.616331	0.860436
U. of Michigan at Ann Arbor	0.629340	0.655179
California Tech	1.000000	1.000000
Washington U. in St. Louis	0.799824	0.894644
U. of Chicago UCTech	0.371733	0.629332

A few takeaways from these results:

- Switching from CRS to VRS will never hurt the efficiency of a DMU. From an optimization perspective, think of this as adding a constraint which will increase or leave unchanged the objective function value from a minimization linear program.
- Another way to view the impact of adding a returns to scale assumption is that this makes it *easier* for a DMU to find some way of being *best* or efficient. They can do this by now having the largest of any output or the smallest of input as well as many other ways.
- At a simple level, the CCR (CRS) efficiency score can be thought of as a combination of how efficient someone is considering their operational and

size efficiency. The BCC (VRS) efficiency is how efficient they are given their current operating size.
- The use of `drop=FALSE` in the subsetting of UTT_CCR.Res enables retaining the original structure of the matrix, which allows the column names to remain. The default is drop=TRUE which simplifies the data object into a form that doesn't have a column name.

5.8 Extracting Multiplier Weights from Sensitivity Analysis

The following process can be used for extracting the shadow prices (row duals) from the linear program results. This extraction is based upon knowing the structure of the model that was built.

```
res.vweight    <- matrix(rep(-1.0, ND*NX), nrow=ND,ncol=NX)
res.uweight    <- matrix(rep(-1.0, ND*NY), nrow=ND,ncol=NY)

Vnames<- lapply(list(rep("v",NX)),paste0,1:NX) # Input weight: v1, ...
Unames<- lapply(list(rep("u",NY)),paste0,1:NY) # Output weight: u1, ...

dimnames(res.vweight)     <- c(DMUnames,Vnames)
dimnames(res.uweight)     <- c(DMUnames,Unames)

rduals <- as.matrix(get_row_duals(res))
res.vweight[k,] <- -1*rduals[1:NX]
   # Extract first NX row duals assuming first constraints are inputs
   # Multiply input weights by -1 to adjust for inequality direction
res.uweight[k,] <- rduals[(NX+1):(NX+NY)]
   # Extract next NY row duals, assuming output constraints follow inputs
weights <- t(c(res.vweight[k,],res.uweight[k,]))
kbl (weights, booktabs=T, digits=5, escape=F,
     col.names=c("$v_{\\text{Research Spending}}$",
                 "$u_{\\text{Licensing Income}}$",
                 "$u_{\\text{Licenses and Options}}$",
                 "$u_{\\text{Startups}}$",
                 "$u_{\\text{Patents Issued}}$"),
     row.names=F,
     caption="Input and Output Weights for Last DMU") |>
  kable_styling(latex_options = "hold_position")
```

TABLE 5.15 Input and Output Weights for Last DMU

$v_{\text{Research Spending}}$	$u_{\text{Licensing Income}}$	$u_{\text{Licenses and Options}}$	u_{Startups}	$u_{\text{Patents Issued}}$
0.00307	0	0.00835	0	0.00342

These results are for just the last unit examined.

Let's now wrap this into the full series of linear programs to calculate the multiplier weights for each unit.

```
RTS<-"CRS"
for (k in 1:ND) {

  modUTT_CRS <- MIPModel() |>
    add_variable(vlambda[j], j = 1:ND, type = "continuous",
                 lb = 0) |>
    add_variable(vtheta, type = "continuous") |>
    set_objective(vtheta, "min") |>
    add_constraint(sum_expr(vlambda[j] * xdata[j,i], j = 1:ND)
                   <= vtheta * xdata[k,i], i = 1:NX,
                   .show_progress_bar=FALSE ) |>
    add_constraint(sum_expr(vlambda[j] * ydata[j,r], j = 1:ND)
                   >= ydata[k,r], r = 1:NY,
                   .show_progress_bar=FALSE )
    if (RTS=="VRS") {modUTT_CRS <-
      add_constraint(modUTT_CRS,
                 sum_expr(vlambda[j],j = 1:ND) == 1) }
    resUTT_CRS <- solve_model(modUTT_CRS,
                              with_ROI(solver = "glpk"))

    res.efficiency[k] <-  get_solution(resUTT_CRS, vtheta)
    res.lambda[k,] <- t(as.matrix(as.numeric(
                     get_solution(resUTT_CRS,
                                  vlambda[j])[,3] )))
    rduals  <- as.matrix(get_row_duals(resUTT_CRS))
    res.vweight[k,] <- -1*rduals[1:NX]
    res.uweight[k,] <- rduals[(NX+1):(NX+NY)]
    }
```

TABLE 5.16 Multiplier Weights from UTT Example

θ^{CRS}	$v_{\text{Research Spending}}$	$u_{\text{Licensing Income}}$	$u_{\text{Licenses and Options}}$	u_{Startups}	$u_{\text{Patents Issued}}$
1.00000	0.00409	0.00528	0.01416	0.00000	0.00000
0.46629	0.00036	0.00040	0.00096	0.00000	0.00065
1.00000	0.00131	0.00000	0.00493	0.00000	0.00000
0.69963	0.00144	0.00101	0.00496	0.02340	0.00000
0.82406	0.00194	0.00136	0.00668	0.03151	0.00000
0.87876	0.00234	0.00164	0.00803	0.03792	0.00000
0.79248	0.00307	0.00215	0.01056	0.04987	0.00000
0.59221	0.00727	0.01227	0.00000	0.00000	0.01904
0.54426	0.00175	0.00123	0.00602	0.02841	0.00000
0.75426	0.00327	0.00411	0.00000	0.08796	0.00000
0.50535	0.00289	0.00203	0.00995	0.04697	0.00000
0.78556	0.00097	0.00000	0.00351	0.01577	0.00000
0.44404	0.00120	0.00084	0.00412	0.01946	0.00000
0.43299	0.00307	0.00215	0.01055	0.04982	0.00000
0.42079	0.00169	0.00119	0.00582	0.02748	0.00000
0.59478	0.00345	0.00242	0.01187	0.05602	0.00000
0.39885	0.00487	0.00823	0.00000	0.00000	0.01277
0.41526	0.00141	0.00099	0.00484	0.02286	0.00000
0.53179	0.00318	0.00358	0.00848	0.00000	0.00577
0.51266	0.00381	0.00267	0.01310	0.06185	0.00000
0.61633	0.00320	0.00000	0.00938	0.00000	0.00578
0.62034	0.00133	0.00000	0.00479	0.02152	0.00000
1.00000	0.00257	0.00000	0.00755	0.00000	0.00465
0.79982	0.00240	0.00000	0.00703	0.00000	0.00434
0.37173	0.00307	0.00346	0.00819	0.00000	0.00558

A few things should be noticed with respect to the multiplier weights:

- Multiplier weights can be quite small because in general, they are multiplied by inputs and the product is typically less than one. They may appear to round to zero but still be significant.
- Multiplier weights are not unique for strongly efficient DMUs. This is explored in more detail in the multiplier chapter of *DEA Using R* and in other DEA specific publications.
- Multiplier weights are usually unique for inefficient DMUs and should match the results obtained using either the row duals of the envelopment model or those obtained directly from the multiplier model.
- People new to DEA are often "offended" by units that place zero weight on certain inputs or on certain outputs as was done by nursing homes B and D. Approaches such as weight restrictions exist to deal with this directly should this be a problem but should be carefully considered.

5.9 Slack Maximization

Situations can arise where units may appear to be radially efficient but can still find opportunities to improve one or more inputs or outputs. This is defined as being weakly efficient.

In order to accommodate this, we need to extend the simple radial model by adding variables to reflect nonradial slacks. We do this by converting the model's input and output constraints from inequalities into equalities by explicitly defining slack variables.

$$\begin{aligned}
&\text{minimize } \theta \\
&\text{subject to } \sum_{j=1}^{N^D} x_{i,j}\lambda_j - \theta x_{i,k} + s^x_i = 0 \; \forall \; i \\
&\sum_{j=1}^{N^D} y_{r,j}\lambda_j - s^y_r = y_{r,k} \; \forall \; r \\
&\lambda_j, s^x_i, s^y_r \geq 0 \; \forall \; j, i, r
\end{aligned}$$

Simply formulating the model with slacks is insufficient. We want to maximize these slacks after having found the best possible radial contraction (minimum value of θ.) This is done by adding a term summing the slacks to the objective function. Note that this sum of slacks is multiplied by ϵ, which is a non-Archimedean infinitesimal.

The value of ϵ should be considered to be so small as to ensure that minimizing theta takes priority or maximizing the sum of slacks. Note also that maximizing the sum of slacks is denoted by minimizing the negative sum of slacks.

$$\begin{aligned}
&\text{minimize } \theta - \epsilon(\sum_i s^x_i + \sum_r s^y_r) \\
&\text{subject to } \sum_{j=1}^{N^D} x_{i,j}\lambda_j - \theta x_{i,k} + s^x_i = 0 \; \forall \; i \\
&\sum_{j=1}^{N^D} y_{r,j}\lambda_j - s^y_r = y_{r,k} \; \forall \; r \\
&\lambda_j, s^x_i, s^y_r \geq 0 \; \forall \; j, i, r
\end{aligned}$$

A common mistake in implementing DEA is to use a finite approximation for ϵ such as 10^{-6}. Any finite value can cause distortions in θ. For example, an application comparing companies using revenue and expenses might have inputs and outputs on the order of millions or billions. In this case, non-radial slacks could also be on the order of 10^6. Multiplying the two results in a value similar in magnitude to the maximum possible efficiency score ($10^{-6} \cdot 10^6 = 1$) which would then potentially overwhelm the radial efficiency, θ, part of the objective function and lead to distorted results.

The proper way to implement slack maximization is to treat it as a preemptive goal programming problem. The primary goal is to minimize θ in a first phase linear program and the second goal, holding the level of θ fixed from the first phase, is to then maximize the sum of the slacks.

The first phase can take the form of any of our earlier linear programs without the ϵ and sum of slacks in the objective function. The second phase is the following where θ^* is the optimal value from phase one and θ is then held constant in the second phase. This is implemented by adding a constraint, $\theta = \theta^*$ to the second phase linear program.

$$
\begin{split}
\text{maximize} \;\; & \sum_i s_i^x + \sum_r s_r^y \\
\text{subject to} \;\; & \sum_{j=1}^{N^D} \lambda_j = 1 \\
& \sum_{j=1}^{N^D} x_{i,j}\lambda_j - \theta x_{i,k} + s_i^x = 0 \; \forall \; i \\
& \sum_{j=1}^{N^D} y_{r,j}\lambda_j - s_r^y = y_{r,k} \; \forall \; r \\
& \theta = \theta^* \\
& \lambda_j, s_i^x, s_r^y \geq 0 \; \forall \; j, i, r
\end{split}
$$

Implementing this is straightforward.

```
RTS<-"CRS"
for (k in 1:ND) {

  LPSlack <- MIPModel()                                          |>
    add_variable(vlambda[j], j = 1:ND, type = "continuous",
```

```
              lb = 0)                                        |>
    add_variable(vtheta, type = "continuous")                |>
    add_variable(xslack[i], i = 1:NX, type = "continuous",
              lb=0)                                          |>
    add_variable(yslack[r], r = 1:NY, type = "continuous",
              lb=0)                                          |>

    set_objective(vtheta, "min")                             |>

    add_constraint(sum_expr(vlambda[j] * xdata[j,i] +
                              xslack[i], j = 1:ND)
                         - vtheta * xdata[k,i]==0, i = 1:NX,
                          .show_progress_bar=FALSE )         |>

    add_constraint(sum_expr(vlambda[j] * ydata[j,r] -
                              yslack[r], j = 1:ND)
                          ==ydata[k,r], r = 1:NY,
                          .show_progress_bar=FALSE )
    if (RTS=="VRS") {LPSlack<-
      add_constraint(LPSlack, sum_expr(vlambda[j],
                                          j = 1:ND) == 1) }
  #Returns to Scale
    result <- solve_model(LPSlack, with_ROI(solver = "glpk"))
    # The following are key steps to slack maximization
    phase1obj <-  get_solution(result, vtheta)
    # Get Phase 1 objective value
    add_constraint(LPSlack, vtheta==phase1obj)
        # Passing result from phase 1 to phase 2

    set_objective(LPSlack, sum_expr(
              xslack[i], i=1:NX)+sum_expr(
                yslack[r], r=1:NY), "max")
          # Modify the objective function for phase 2
    result <- solve_model(LPSlack, with_ROI(solver = "glpk"))

    res.efficiency[k] <-  get_solution(result, vtheta)
    res.lambda[k,] <- t(as.matrix(
              as.numeric(get_solution(result,
                                        vlambda[j])[,3] )))
}
```

```
kbl(head(cbind(res.efficiency,poscol(res.lambda[1:25,]))),
    booktabs=T, escape=F,
    caption="Selected Input-oriented Efficiency Results
    with Slack Maximization") |>
  kable_styling(latex_options = "hold_position")
```

5.10 DEA Packages

Over the years, there have been several R packages developed for doing DEA. Each package is often developed by a researcher or team for their specific needs. Some continue to be enhanced over time. At this point the reader will have an understanding of many of the basics of the implementation and have a greater appreciation of how the packages. The following packages are currently available through CRAN.

- `Benchmarking` by Bogetoft and Otto stands out for the comprehensive toolset and documentation in their book.
- `deaR` provides some interesting classic datasets along with both standard and less commonly used variations of DEA.
- `DJL` incorporates a variety of DEA techniques and is officially an abbreviation of distance measurement based judgement and learning. Perhaps not too coincidentally, it matches the author's, Dong-Joon Lim's, initials.
- `MultiplierDEA` was developed by Aurobindh Kalathil Kumar from Portland State University and emphasizes the multiplier model of DEA.
- `nonparaeff` draws its name from the nonparametric nature of DEA.
- `rDEA` stands for Robust DEA.

In each case, the notation for calling DEA varies. This includes the way that data is structured, how returns to scale is specified, whether a second phase for slack maximization is conducted, and more. The result is that reading the help file for a package is necessary.

5.11 DEA Model Building

"All models are wrong but some are useful" – George Box, 1978

5.11.1 Selection of Inputs and Outputs

At its simplest, the analyst should try to build a model that reflects the most important resources used in the production of the unit's most important outputs. No DEA input-model is perfect and they all represent tradeoffs between data availability, accuracy, and other issues.

The subject of DEA model building is involved and arguably as much an art as it is a science. Economists will argue (correctly) that the production function axioms that underlie theoretical foundations of DEA require that inputs can have tradeoffs between them in the production of the outputs and that outputs also have tradeoffs between them. Also, that all inputs contribute to all outputs. In other words, a benchmarking study of hospitals that had as one one of the inputs heart surgeons and as one of the outputs heart transplants performed but another output such as kidney transplants could be problematic.

Data Envelopment Analysis is by its very name data driven. This highlights the fact that data is needed for conducting the analysis. Unlike multiple regression in which the need for more observations as a function of the number of independent variables is well known, there is no simple sufficiency test for DEA models. Common heuristics of needing at least three times as many DMUs as the input-output model complexity is generally helpful. (Note that model complexity is sometimes interpreted as number of inputs plus the number of outputs or as the product of the number of inputs and outputs.)

Since DMUs can be efficient by being the best on any ratio of an output to an input, the higher the number of inputs and outputs, the less *discrimination* between DMUs is typically found. This *curse of dimensionality* means that highly complex models may find most or all DMUs to be efficient. Forced fixes to adjust for this often lead to difficulties.

Correlations among inputs and correlations among outputs do not generally affect the efficiency score. In effect, two highly correlated inputs tend to behave like one and do not increase the *curse of dimensionality* although they might have unintended consequences elsewhere and violate production economics ideas. Also, they may cause multiplier weights to be less interpretable.

Note that if an input or an output is inaccurate, this affects the accuracy of the final efficiency scores.

Furthermore, DEA is an extreme point technique. As such, it is important that the frontier points be valid instances. A bad data point (DMU) can have a profound affect on the efficiency of many other units. As such, the analyst should carefully review the efficient points. (Techniques such as super-efficiency can be used to highlight and prioritize efficient DMUs worth detailed examination.)

5.11.2 Model Choices

The issues of returns to scale and input vs. output-orientation should be carefully justified in terms of the particular application being examined.

With respect to returns to scale, it is straightforward to examine results and consider what makes the most sense in the application. Does it really make sense to scale up or down a unit to an extreme extent without distorting their actual operations?

In terms of orientation, is the primary goal to increase outputs or decrease inputs? Note that there are DEA models that allow for both simultaneously but input and output orientation remain the most commonly used orientations.

5.11.3 Application Area Expertise

It is always critical to bring solid application expertise to any DEA study in all phases of model building and at the interpretation stage as well. Years ago, I did a study of university technology transfer performance and found the efficient universities included schools that might be expected such as MIT and Georgia Tech but I found the inclusion of BYU surprising. I went and talked with Julie Reed, a lawyer specializing intellectual proper and technology transfer. She said that she was not at all surprised to see BYU listed as efficient – they are small but very highly respected for their technology transfer. While not a formal validation, this is an indirect form of validation.

There is no detailed statistical test for the validity of a DEA model and results. This means that there is greater responsibility on the analyst to ensure that a meaningful model was built and appropriate model options were used.

Lastly, it is always helpful to look to the literature. With over 10,000 published DEA papers since 1977, you can almost always find a paper in the same or similar application area. Look at the authors' choices of inputs and output as well as the particulars of the DEA model being used. This can serve as a starting point for building an even better paper. Having said that, just because a paper or presentation uses a particular input-output model doesn't mean that it is a good model. A recent presentation at a major international conference highlights that previous studies are not always great examples. The authors conducted a benchmarking study of hospitals. The inputs were relatively traditional such as number of nurses and hospital beds. Outputs included traditional measures such as the number of patients served but one output was unique, *number of deaths* – yes, researcher based his analysis on the implicit assumption that the route to efficiency for inefficient hospitals was to *increase the number of deaths!* While the research stated that this was a measure of workload for the hospital, this clearly a bad modeling choice. There are DEA models that can accommodate "bad outputs" such as toxic byproducts in manufacturing or energy product. Another option would be to emphasize the lives saved.

In short, a DEA application should always bring together both application area expertise and familiarity with DEA to ensure meaningful results.

5.12 Further Reading

This section covers the second chapter of the book and introduces DEA and R working through a model and looping through a series of analyses. The interested reader is referred to the book *DEA Using R* for more information on DEA. In particular, chapters covers topics such as:

- a matrix-oriented implementation of DEA,
- output-orientation (maximizing the production of outputs)
- multiplier model (weighting inputs and weighting models)
- an easy to understand application applied to over 100 years of data
- examples of previously published applications applied to R&D project evaluation.

5.13 Exercises

Exercise 5.1 (Graphically Adding DMU E). Add a fifth unit, E, to the first example that produces 400 units output using 30 units of input. Graphically evaluate all five units for their efficiency scores and lambda values. Interpret the solution in terms of who is doing well, who is doing poorly, and who should be learning from whom.

Exercise 5.2 (Adding DMU E Using R). Examine the new unit, E, using R. Interpret the solution in terms of who is doing well, who is doing poorly, and who should be learning from whom.

Exercise 5.3 (Looping). Wrap a for loop around the model from the previous exercise to examine every unit. Discuss results.

Exercise 5.4 (Bigger Data). Use a bigger data set and conduct an analysis & interpretation (more inputs, outputs, and units.)

Exercise 5.5 (Compare Results). Check results against a DEA package (ex. `DEAMultiplier`, `Benchmarking`, `nonparaeff`).

Exercise 5.6 (Slacking). Construct an example where Phase 2 increases positive slacks from Phase 1.

Exercise 5.7 (DEA Application). Conduct a DEA study for an application that you are personally familiar with. (Pick one for which data is readily available but something that you are passionate about. It can have scrubbed or anonymized data. Examples might include your favorite sports league, team salaries, coaching salaries, wins, etc. or it can be USAF related.)

a. Describe the application.
b. Describe and justify the data including the inputs and outputs used as well as items explicitly not used.
c. Select an appropriate DEA model and conduct the analysis.
d. Discuss the results.

Exercise 5.8 (DEA Technique Choices). Choosing Inputs and Outputs: The DEA store example in chapter 5 uses an input-output model of sales and employees. Describe a richer input-output model that you feel would make for a better case. Be sure to discuss tradeoffs and other DEA model choices, including how results could be used.

Exercise 5.9 (Using DEA to Compare Products). Conduct a DEA study on the given sample data about cameras specifications in different smartphone models from different companies i.e. Apple, Samsung, Google, LG etc. The inputs can be camera characters and output can be cost of phone. See an example below:

TABLE 5.17 Camera Data

	Cost($)	Storage Space(GB)	Resolution(MP)	Lens Thickness(mm)
Phone A	250	16	12	4
Phone B	225	16	8	5
Phone C	300	32	16	4.5
Phone D	275	32	8	4

a. Add more DMUs to above data so that there are minimum 10 in total.

b. Add more inputs and outputs as needed. Describe and justify the data including the inputs and outputs used as well as items explicitly not used.
c. Select an appropriate DEA model and conduct the analysis.
d. Discuss the results.

6

Mixed Integer Optimization

Up until this point, we have always assumed that variables could take on fractional (or non-integer) values. In other words, they were continuous variables. Sometimes the values conveniently turned out to be integer-valued, in other cases, we would brush off the non-integer values of the results.

Allowing for integer values opens up many more important areas of applications as we will see. Let's look at two numerical examples. The first shows a case where integrality makes little difference, the second where it has a major impact.

Unfortunately, there is no such thing as a free lunch-adding and integrality can make problems much more computationally demanding. We will go through an example of how many optimization routines do integer optimization to demonstrate why it can be more difficult algorithmically even if looks trivially easy from the perspective of the change in the `ompr`function.

6.1 Example of Minor Integrality Impact

Let's revisit the earlier example of producing drones with one additional hour in assembly as we did in Chapter 4. As a reminder, the LP is the same three variable case we have been using with a change of 901 instead of 900 for the second constraint's right-hand side.

$$\begin{split}
\text{Max } & 7 \cdot Ants + 12 \cdot Bats + 5 \cdot Cats \\
\text{s.t.: } & 1 \cdot Ants + 4 \cdot Bats + 2 \cdot Cats \leq 800 \\
& 3 \cdot Ants + 6 \cdot Bats + 2 \cdot Cats \leq 901 \\
& 2 \cdot Ants + 2 \cdot Bats + 1 \cdot Cats \leq 480 \\
& 2 \cdot Ants + 10 \cdot Bats + 2 \cdot Cats \leq 1200 \\
& Ants, Bats, Cats \geq 0
\end{split}$$

We use the following implementation.

```
model1 <- MIPModel() |>
  add_variable(Ants, type = "continuous", lb = 0) |>
  add_variable(Bats, type = "continuous", lb = 0) |>
  add_variable(Cats, type = "continuous", lb = 0) |>

  set_objective(7*Ants + 12*Bats + 5*Cats,"max")  |>

  add_constraint(1*Ants + 4*Bats + 2*Cats<=800)   |> # machining
  add_constraint(3*Ants + 6*Bats + 2*Cats<=901)   |> # assembly
  add_constraint(2*Ants + 2*Bats + 1*Cats<=480)   |> # testing
  add_constraint(2*Ants + 10*Bats + 2*Cats<=1200)    # sensors

result1 <-  solve_model(model1, with_ROI(solver="glpk"))

inc_as_res <- cbind (result1$objective_value,
                     t(result1$solution))
colnames(inc_as_res)<-c("Profit", "Ants", "Bats", "Cats")
kbl (inc_as_res, digits=6, booktabs=T,
     caption="Production Plan with Continuous Variables") |>
  kable_styling(latex_options = "hold_position")
```

TABLE 6.1 Production Plan with Continuous Variables

Profit	Ants	Bats	Cats
2227.25	50.5	0	374.75

In this case, the fractional value of *Cats* would be only somewhat concerning. We wouldn't actually ship an 80% complete Cat drone to a customer-at least I hope not. The difference between 374 and 375 is relatively small over a month of production. This is a difference of much less than 1% and none of the numbers specified in the model appear to be reported to more than two significant digits or this level of precision. The result is that we could specify the answer as 374.8 and be satisfied that, while it is our actual best guess or might reflect a chair half finished for the next month, it isn't a major concern.

For the sake of illustration, let's show how we would modify the model to be integers. If we had wanted to modify the problem to force the amount of each item to be produced to be an integer value, we can modify our formulation and the implementation with only a few small changes.

In the formulation, we can replace the non-negativity requirement with a restriction that the variables are drawn from the set of non-negative integers.

$$
\begin{aligned}
\text{Max } & 7 \cdot Ants + 12 \cdot Bats + 5 \cdot Cats \\
\text{s.t.: } & 1 \cdot Ants + 4 \cdot Bats + 2 \cdot Cats \leq 800 \\
& 3 \cdot Ants + 6 \cdot Bats + 2 \cdot Cats \leq 901 \\
& 2 \cdot Ants + 2 \cdot Bats + 1 \cdot Cats \leq 480 \\
& 2 \cdot Ants + 10 \cdot Bats + 2 \cdot Cats \leq 1200 \\
& Ants,\ Bats,\ Cats \in \{0, 1, 2, 3, ...\}
\end{aligned}
$$

The `ompr` implementation has a similar, straightforward change. In the declaration of variables (`add_variable` function), we simply set the type to be `integer` rather than `continuous`. Since nothing else is changing in the model, we can simply re-add the variables to redefine them from being continuous to be integer.[1]

```
model1 <- add_variable(model1, Ants, type = "integer", lb = 0)
model1 <- add_variable(model1, Bats, type = "integer", lb = 0)
model1 <- add_variable(model1, Cats, type = "integer", lb = 0)
```

Let's confirm that the implemented model is of the form desired.

```
model1
```

```
## Mixed integer linear optimization problem
## Variables:
##   Continuous: 0
##   Integer: 3
##   Binary: 0
## Model sense: maximize
## Constraints: 4
```

We can see that it now has three integer variables replacing the variables of the same name that were continuous so let's move on to solving it and comparing our results.

```
result1 <-  solve_model(model1, with_ROI(solver="glpk"))
```

[1]ompr 1.0 adds an enforced variable checking in the model building. This now requires rebuilding the model rather than just redefining the same named variable.

TABLE 6.2 Production Plans Based on Variable Type

	Profit	Ants	Bats	Cats
Continuous	2227.25	50.5	0	374.75
Integer	2227.00	51.0	0	374.00

Notice that in this case, there was a small adjustment to the production plan and a small decrease to the optimal objective function value. This is not always the case, sometimes the result can be unchanged. In other cases, it may be changed in a very large way. In still others, the problem may in fact even become infeasible.

6.2 Example of Major Integality Impact

Let's take a look at another example. Acme makes two products, let's refer to them as product 1 and 2. Product 1 generates a profit of $2000 per product, requires three liters of surfactant for cleaning, and eight kilograms of high grade steel.

In contrast, each unit of Product 2 produced generates a higher profit of $3000 per product, requires eight liters of surfactant, and three kilograms of high grade steel. Acme has 31 liters of surfactant and 26 kilograms of high grade steel.

Only completed products are sellable, what should be Acme's production plan?

Formulating the optimization problem is similar to our earliest production planning problems. Let's define x_1 to be the amount of product 1 to produce and x_2 to be the amount of product 2 to produce.

$$
\begin{aligned}
\text{Max } & 2x_1 + 3x_2 \\
\text{s.t. } & 3x_1 + 9x_2 \leq 31 && \text{[Surfactant]} \\
& 8x_1 + 3x_2 \leq 26 && \text{[Steel]} \\
& x_1,\ x_2 \in \{0, 1, 2, 3, ...\}
\end{aligned}
$$

At first glance, this looks the same as our earlier drone eample but the last constraint cannot be solved directly strictly using linear programming. It is instead referred to as an integer linear program or ILP.

Again, we could try solving it with and without the imposition of the integrality constraint.

```
LPRmod <- MIPModel() |>
  add_variable(Vx1, type = "continuous", lb = 0) |>
  add_variable(Vx2, type = "continuous", lb = 0) |>
  set_objective(2*Vx1 + 3*Vx2, "max")          |>
  add_constraint(3.0*Vx1 + 9.0*Vx2 <= 31)      |> # Surfactant
  add_constraint(8.0*Vx1 + 3.0*Vx2 <= 26)         # Steel

LPSolR <-  solve_model(LPRmod, with_ROI(solver = "glpk"))

obj_val <- objective_value(LPSolR)
x1 <- get_solution (LPSolR, 'Vx1')
x2 <- get_solution (LPSolR, 'Vx2')
acme_res_lp  <- cbind(x1,x2,obj_val)
colnames(acme_res_lp) <- list("x1", "x2", "Profit")
rownames(acme_res_lp) <- "LP Solution"
```

All we need to do is repeat the same process with changing the variable type to `integer` instead of `continuous`.

```
IPmod <- add_variable(LPRmod, Vx1, type = "integer", lb = 0) |>
  add_variable(Vx2, type = "integer", lb = 0)
IPSolR <-  solve_model (IPmod, with_ROI(solver = "glpk"))
obj_val <- objective_value(IPSolR)
```

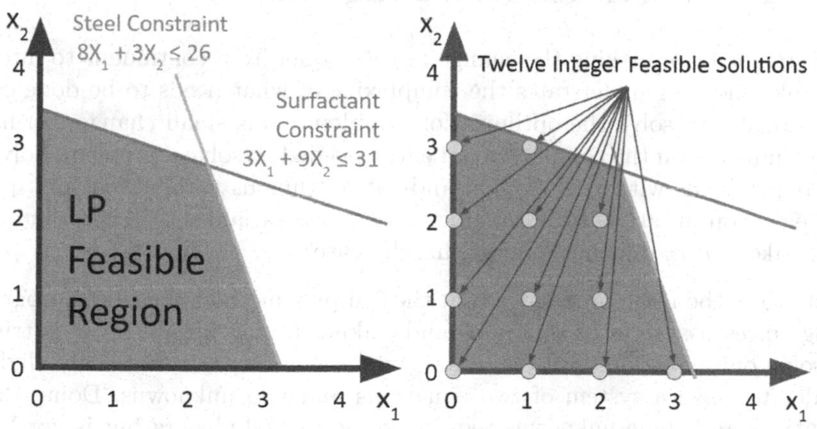

FIGURE 6.1 LP Region and 12 IP Feasible Solutions.

```
x1 <- get_solution (IPSolR, 'Vx1')
x2 <- get_solution (IPSolR, 'Vx2')
acme_res_ip  <- cbind(x1,x2,obj_val)

colnames(acme_res_ip) <- list("x1", "x2", "Profit")
rownames(acme_res_ip) <- "IP Solution"
kbl (rbind(acme_res_lp,acme_res_ip), booktabs=T, digits=3,
     caption="Production Plan-Continuous vs. Integer") |>
  kable_styling(latex_options = "hold_position")
```

TABLE 6.3 Production Plan-Continuous vs. Integer

	x1	x2	Profit
LP Solution	2.238	2.698	12.571
IP Solution	1.000	3.000	11.000

Simply rounding down did not yield an optimal solution. Notice that production of product 1 went down while product 2 went up. The net impact was a profit reduction of almost *10%*.

6.3 The Branch and Bound Algorithm

The simplicity of making the change in the model from continuous to integer variables hides or understates the complexity of what needs to be done computationally to solve the optimization problem. This small change can have major impacts on the computational effort needed to solve a problem. Solving linear programs with tens of thousands of continuous variables is very quick and easy computationally. Changing those same variables to general integers can make it *very* difficult. Why is this the case?

At its core, the basic problem is that the Simplex method and linear programming solves a system of equations and unknowns (or variables). It is trivial to solve one equation with one unknown such as $5x = 7$. It is only slightly harder to solve a system of two equations and two unknowns. Doing three equations and three unknowns requires some careful algebra but is not hard. In general, solving n equations and n unknowns is readily solvable. On the other hand, none of this algebraic work of solving for unknown variables can require that the variables take on integer values. Even with only one equation (say $C \cdot x = 12$), one unknown variable (x), and one constant (C), x will only be integer for certain combinations of numbers in the above equation. It gets even less likely to have an all integer solution as the number of equations and unknowns increase.

Another way to visualize integer solutions for general systems of inequalities is to think of the case with two equations and two unknowns (x and y). Finding a solution is the same as finding where two lines intersect in two-dimensional space. The odds of getting lucky and having the two lines intersect at an integer solution is small. Even if we limit ourselves to x and y ranging from 0 to 10, this means that there are $11^2 = 121$ possible integer valued points in this space such as (0,0), (0,3), and (10,10). In contrast, the set of all points, including non-integers, in this region includes all of the above points as well as (0,0.347) and (4.712, 7.891) among infinitely more. The result is that if you think of this as a target shooting, the probability of finding an integer solution by blind luck is $\frac{121}{\infty} \approx 0$.

What we need is an algorithm around the Simplex method to accomplish this. A basic and common and approach often used for solving integer variable linear programming problems is called Branch and Bound. We start with a solution that has presumably continuous valued variables. We then branch (create subproblems) off of this result for variables that have fractional values but should be integers and add bounds (additional constraints) to rule out the fractional value of that variable.

Given the importance of integer programming and the impact on solving speed, we'll demonstrate the process by showing the steps graphically and in R.

6.3.1 The LP Relaxation

To solve this using branch and bound, we *relax* the integrality requirements and solve what is called the LP Relaxation. To do this, we simply remove the integrality requirement.

$$\begin{aligned}
\text{Max } & 2x_1 + 3x_2 \\
\text{s.t. } & 3x_1 + 9x_2 \leq 31 \quad \text{[Surfactant]} \\
& 8x_1 + 3x_2 \leq 26 \quad \text{[Steel]} \\
& x_1,\ x_2 \geq 0
\end{aligned}$$

Unlike requiring the variables to be integer valued, this is a problem that we **can** solve using linear programming so let's go ahead do it! Well, we already did so let's just review the results for what is now called the LP Relaxation because we are *relaxing* the integrality requirements.

```
kbl (acme_res_lp, booktabs=T, digits=3,
     caption="Acme's Production Plan based on the LP Relaxation") |>
  kable_styling(latex_options = "hold_position")
```

TABLE 6.4 Acme's Production Plan Based on the LP Relaxation

	x1	x2	Profit
LP Solution	2.238	2.698	12.571

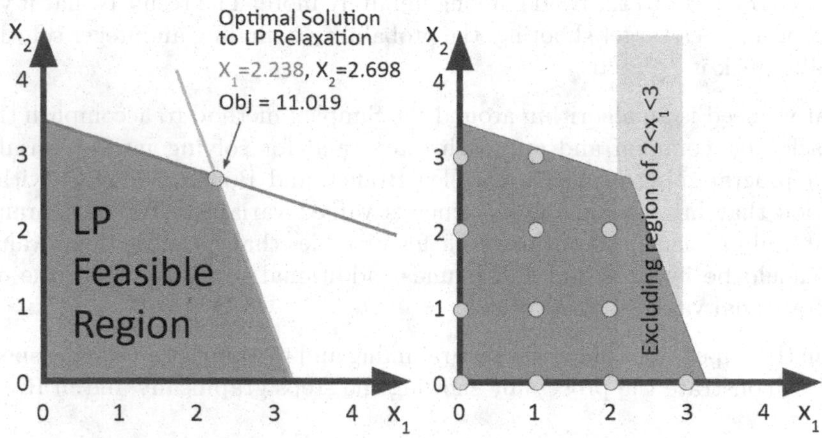

FIGURE 6.2 LP Relaxation.

6.3.2 Subproblem I

Alas, this production plan from the LP Relaxation is not feasible from the perspective of the original integer problem because it produces *2.238* of product 1 and *2.698* of product 2. If both of these variables had been integer we could have declared victory and been satisfied that we had easily (*luckily?*) found the optimal solution to the original integer programming problem so quickly.

Instead, we will need to proceed with the branch and bound process. Since both of the variables in the LP relaxation have fractional values, we need to start by choosing which variable to use for branching first. Algorithmic researchers would focus on how to best pick a branch but for our purposes to improve solution speed, but it doesn't really matter for illustration so let's arbitrarily choose to branch on x_1.

For the branch and bound algorithm, we want to include two subproblems that exclude the *illegal* value of $x_1 = 2.238$ as well as everything else between 2 and 3. We say that we are *branching* on x_1 to create two subproblems. The first subproblem (I) has an added constraint of $x_1 \leq 2.0$ and the other subproblem (II) has an added constraint of $x_1 \geq 3.0$.

$$\begin{aligned} \text{Max } & 2x_1 + 3x_2 \\ \text{subject to } & 3x_1 + 9x_2 \leq 31 \\ & 8x_1 + 3x_2 \leq 26 \\ & x_1 \leq 2.0 \qquad \text{[Bound for Subproblem I]} \\ & x_1,\ x_2 \geq 0 \end{aligned}$$

Since this is just a bound on the decision variable, we can implement it by just modifying the LP Relaxation's model `LPRmod` that we created earlier by redefining the x_1 variable (`Vx1`).

```
LPSubI <- add_variable(LPRmod, Vx1, type = "continuous",
                       lb = 0, ub=2.0)
LPSubI <- solve_model(LPSubI, with_ROI(solver = "glpk"))

obj_val <- objective_value(LPSubI)
x1 <- get_solution (LPSubI, 'Vx1')
x2 <- get_solution (LPSubI, 'Vx2')

LPSubI_res  <- cbind(x1,x2,obj_val)
```

TABLE 6.5 Production Plan Based on Subproblem I

x1	x2	obj_val
2	2.778	12.333

Looking over the results, we now get an integer value for x_1 but not for x_2. We repeat the same process by creating subproblems from Subproblem I by branching off of x_2.

6.3.3 Subproblem III

Choosing which subproblem to examine next is one of the areas that large scale integer programming software and algorithms specialize in and provide options. One way to think of it is to focus on searching down a branch and

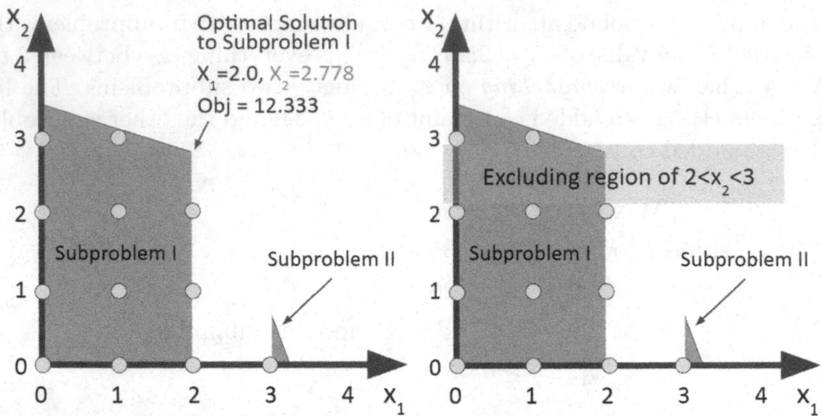

FIGURE 6.3 Subproblem I.

bound tree deeply or to search across the breadth of the tree. For this example, let's go deep which means jumping ahead to Subproblem III but we'll return to Subproblem II later.

Since x_2 is now a non-integer solution, we will create branches with bounds (or constraints) on x_2 in the same manner as before. Subproblem III has an added constraint of $x_2 \leq 2.0$ and Subproblem IV has $x_2 \geq 3.0$.

To simplify the implementation, I can simply change the upper and lower bounds on variables rather than adding separate variables.

```
LPSubIII <- LPRmod |>
  add_variable(Vx1, type = "continuous", lb = 0, ub = 2) |>
  add_variable(Vx2, type = "continuous", lb = 0, ub = 2) |>
  solve_model(with_ROI(solver = "glpk"))
```

TABLE 6.6 Integer Valued Production Plan Based on Subproblem III

x1	x2	obj_val
2	2	10

This results in integer values for both x_1 and x_2 so it is feasible with respect to integrality in addition to of course satisfying, the production constraints. It does generate less profit than the LP relaxation. While it is feasible, it doesn't prove that it is optimal though. We need to explore the other potential branches.

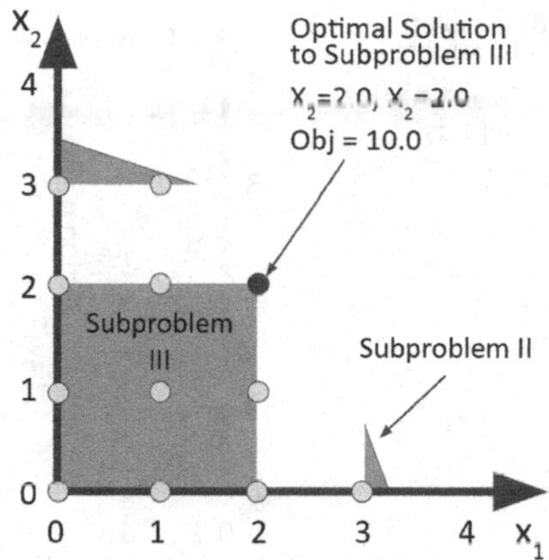

FIGURE 6.4 Subproblem III – An IP Feasible Solution.

6.3.4 Subproblem IV

Next, let's look at Subproblem IV. This problem adds the bound of $x_2 \geq 3.0$ to Subproblem I. Notice that in the `MIPModel` implementation the variable for x_1 (`Vx1`) has an upperbound of 2, (`ub=2`) in order to implement the bounding constraint for Subproblem I and the lower bound on x_2 of 3 in the variable declarations. Notice from the earlier graphical examples, there is no feasible region (blue area) that has $x_2 \geq 3$. By visual inspection, we expect this to be an infeasible subproblem but let's confirm it.

```
LPSubIV <- add_variable(LPRmod, Vx1, type = "continuous",
                        lb = 0, ub = 2)
LPSubIV <- add_variable(LPSubIV, Vx2, type = "continuous",lb = 3)
LPSubIV <-  solve_model(LPSubIV, with_ROI(solver = "glpk"))
```

TABLE 6.7 Production Plan Based on Subproblem IV

x1	x2	obj_val
1.333	3	11.667

While x_2 is now integer valued, $x_1 = 1.333$, so it is no longer integer valued. We need to repeat the process of branching on x_1 again for Subproblems V and VI.

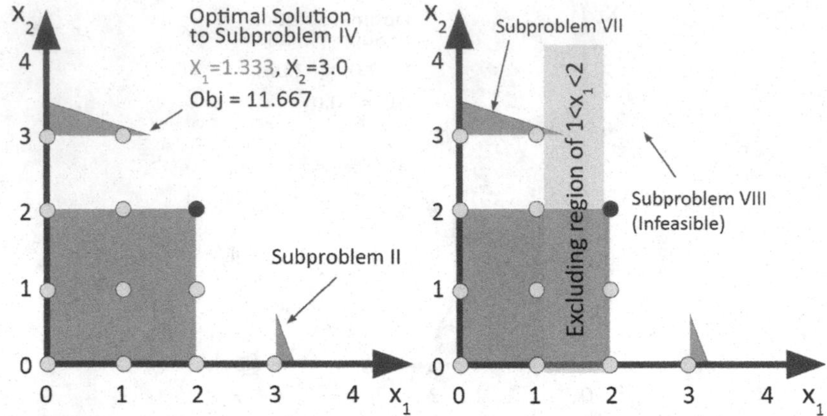

FIGURE 6.5 Subproblem IV.

6.3.5 Subproblem V

For Subproblem V, use a constraint of $x_1 \leq 1$. We'll continue with adjusting the bounds of the variables to do the branch and bound.

```
LPSubV <- add_variable(LPRmod, Vx1, type = "continuous",
                      lb = 0, ub = 1)
LPSubV <- add_variable(LPSubV, Vx2, type = "continuous",lb = 3)
LPSubV <-  solve_model(LPSubV, with_ROI(solver = "glpk"))
```

TABLE 6.8 Production Plan Based on Subproblem V

x1	x2	obj_val
1	3.111	11.333

Unfortunately the solution to Subproblem V now has $x_2 = 3.111$ so we will need to repeat this branch and bound for Subproblems VII and VIII. Before we do that, let's consider Subproblem VI.

6.3.6 Subproblem VI

In this case, we our retaining our previous bounds from Subproblems I and IV while adding a bound for x_1. The result is that our bounds on x_1 is $2 \leq x_1 \leq 2$ or in other words, $x_1 = 2$ while $x_2 \geq 3$. A careful look at the previous

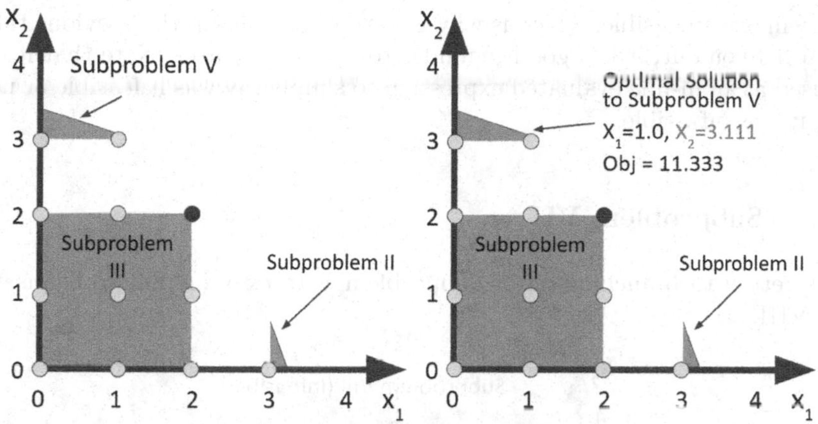

FIGURE 6.6 Subproblems V and VI.

figure suggest there is no feasible solution in this situation but let's solve the corresponding LP to confirm this.

```
LPSubVI <- add_variable(LPRmod, Vx1, type = "continuous",
                        lb = 2, ub = 2)
LPSubVI <- add_variable(LPSubVI, Vx2, type = "continuous",
                        lb = 3)
LPSubVI <- solve_model(LPSubVI, with_ROI(solver = "glpk"))
LPSubVI$status
```

```
## [1] "infeasible"
```

```
obj_val <- objective_value(LPSubVI )
x1 <- get_solution (LPSubVI , 'Vx1')
x2 <- get_solution (LPSubVI , 'Vx2')
LPSubVI_res  <- cbind(x1,x2,obj_val)
```

TABLE 6.9 Infeasible Production Plan Based on Subproblem VI

x1	x2	obj_val
2	3	13

The `ompr` status value of this solved LP indicates that Subproblem VI is *infeasible.* It still returned values that were used when it determined that the problem was infeasible, which is why it gave the results in the previous table. From here on out, it is a good reminder to check the status. Note that it can be used as an in-line evaluated expression to simply say was it feasible or not? Yes, it was infeasible.

6.3.7 Subproblem VII

Let's return to branching off of Subproblem V to examine Subproblems VII and VIII.

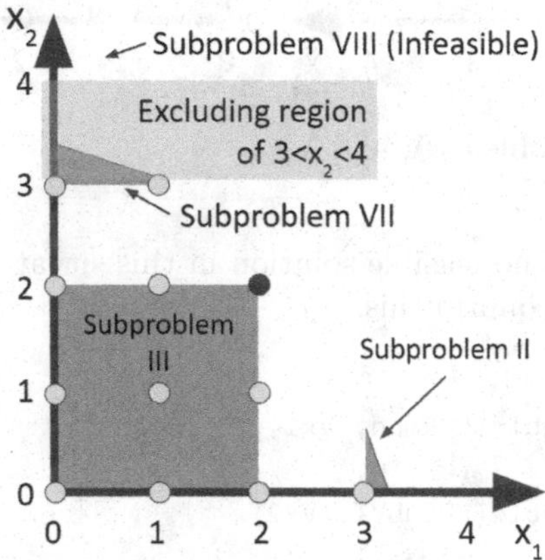

FIGURE 6.7 Subproblem VII and VIII.

```
LPSubVII <- add_variable(LPRmod, Vx1, type = "continuous",
                         lb = 0, ub = 1)
LPSubVII <- add_variable(LPSubVII, Vx2, type = "continuous",
                         lb=0, ub=3)
LPSubVII <-  solve_model(LPSubVII, with_ROI(solver = "glpk"))
```

TABLE 6.10 Integer Valued Production Plan Based on Subproblem VII

x1	x2	obj_val
1	3	11

This solution is all integer valued and the objective function is better than our previous integer valued solution. The result is that we have a new candidate optimal solution. Since there are still branches to explore, we can't declare victory.

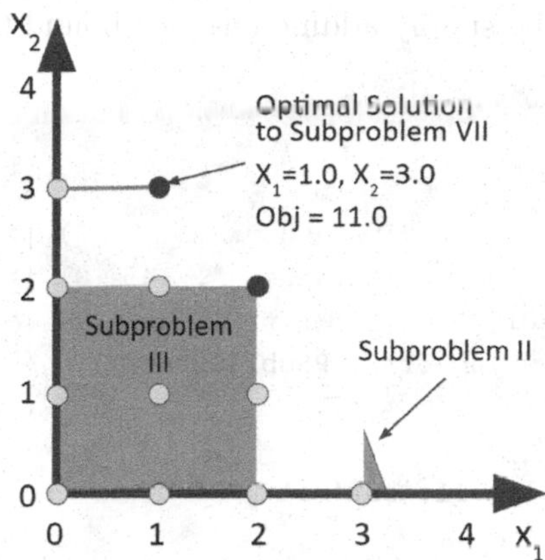

FIGURE 6.8 Subproblem VII Solution.

6.3.8 Subproblem VIII

Setting a lower bound of $x_2 \geq 4$ makes the problem infeasible. We can see from the drawings of the feasible region that the feasible region does not extend to that height at all regardless of what x_1 is. We can confirm by solving Subproblem VIII though since we don't generally have the luxury of looking at the feasible region. Again, the results table highlights that care must be taken to not give credence to values returned from an infeasible solution.

```
LPSubVIII <- add_variable(LPRmod, Vx1, type = "continuous", lb = 1,)
LPSubVIII <- add_variable(LPSubVIII, Vx2, type = "continuous", lb = 4)
LPSubVIII <- solve_model(LPSubVIII, with_ROI(solver = "glpk"))
LPSubVIII$status
```

```
## [1] "infeasible"
```

TABLE 6.11 Infeasible Production Plan Based on Subproblem VIII

x1	x2	obj_val
1	4	14

6.3.9 Subproblem II

Now, we have searched down all branches or subproblems descending from Subproblem I so we can return to Subproblem II. Let's go back and do that problem. We do that by simply adding one new bound to the LP Relaxation. That is $x_1 \geq 3.0$.

```
LPSubII <- LPRmod |>
  add_variable(Vx1, type = "continuous", lb = 3) |>
  add_variable(Vx2, type = "continuous", lb = 0) |>
  solve_model(with_ROI(solver = "glpk"))
cat("Status of Subproblem II:",LPSubII$status)
```

```
## Status of Subproblem II: optimal
```

```
rownames(LPSubII_res) <- ""
```

TABLE 6.12 Production Plan Based on LP Subproblem II

	x1	x2	obj_val
Vx1	3	0.667	8

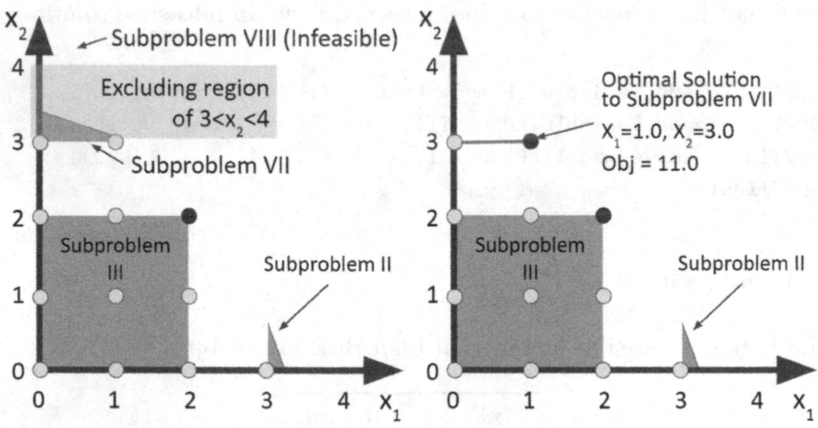

FIGURE 6.9 Subproblems VI and VII.

At this point, our first reaction may be to breathe deeply and do the same branch and bound off of x_2. On the other hand, if we step back and take note that the objective function value is 10.0, while optimal for Subproblem II, it is less than what we found from a feasible, fully integer-valued solution to Subproblem III and the even better Subproblem VII. Given that adding

TABLE 6.13 Acme's Integer Optimal Production Plan Based on Subproblem VII

x1	x2	obj_val
1	3	11

constraints can't improve an objective function value, we can safely trim all branches below Subproblem II.

Given that we no longer have any branches to explore, we can declare that we have found the optimal solution. The optimal solution can now be definitively stated to be what we found from Subproblem VII.

Before this, we could only say that it was feasible solution and candidate to be optimal since no better integer feasible solution had been found.

Let's summarize the results of the series of LPs solved.

TABLE 6.14 Branch and Bound's Sequence of LPs

	x1	x2	Profit
LP Relaxation	2.238	2.698	12.571
Subproblem I	2.000	2.778	12.333
Subproblem III	2.000	2.000	10.000
Subproblem IV	1.333	3.000	11.667
Subproblem V	1.000	3.111	11.333
Subproblem VI	–	–	Infeasible
Subproblem VII	1.000	3.000	11.000
Subproblem VIII	–	–	Infeasible
Subproblem II	3.000	0.667	8.000

The Branch and Bound process is often characterized as a tree. Each subproblem is a circle and represents a separate linear program with its objective function value and decision variable values. The arcs or branches connecting to a new node show the added constraint.

6.4 Computational Complexity

As for computational complexity, this was a small problem with only two integer variables, but it still required solving the LP Relaxation and four separate LP subproblems for a total of five linear programs in all. As the number of variables and constraints increases, so do the number of LPs needed

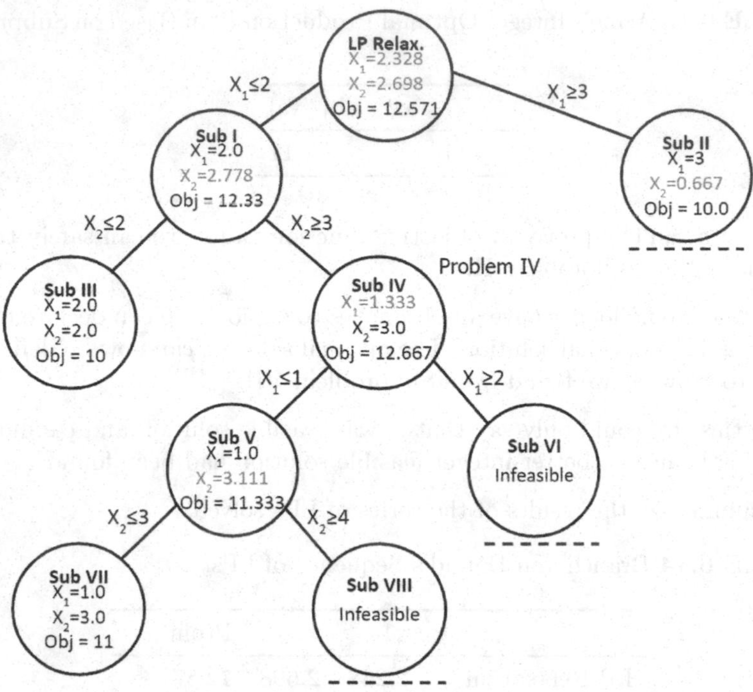

FIGURE 6.10 Branch and Bound Tree.

to typically find a solution using branch and bound. Small and medium size problems can generally be solved quickly but worst case scenarios have a combinatorial explosion.

6.4.1 Full Enumeration

Another approach to integer programming is full enumeration which means listing out all possible solutions and then determining if that solution is feasible and if so, what the objective function value is. Alas, this can result in a combinatorial explosion. For example, if a problem has 1000 non-negative integer variables each of which can range from zero to nine, full enumeration would require listing 10^{1000} possible solutions. This is far larger than the number of grains of sand on Earth (~10^{19}) or the number of stars in the universe (~10^{19}). In fact, this is much larger than the number of atoms in the universe (~10^{80}). Needless to say explicit enumeration for large problems is not an option.

The Branch and Bound algorithm has the benefit that it will find the optimal solution. Sometimes it will find it quickly, other times it may take a very long time.

Worst case scenarios for Branch and Bound may approach that of full enumeration but in practice performs much better. A variety of sophisticated algorithmic extensions have been added by researchers over the years but solving large scale integer programming problems can still be quite difficult.

One option to deal with long solving times is to set early termination conditions such as running for a certain number of seconds. If it is terminated early, it may report the best feasible solution found so far and a precise bound as to how much better an as yet unfound solution might be based on the best remaining open branch. This difference between the best feasible solution found so far and the theoretically possible best solution to be found gives a gap that an analyst can set. This acceptable gap is sometimes referred to as a suboptimality tolerance factor. In other words, what percentage of suboptimality is the analyst willing to accept for getting a solution more quickly. For example, a suboptimality tolerance factor of 10% would tell the software to terminate the branch and bound algorithm if a feasible solution is found, and it is certain that regardless of how much more time is spent solving, it is impossible to improve upon this solution by more than 10%.

In practice, even small suboptimality tolerance factors like 1% can allow big problems to be solved quickly and are often well within the margin of error for model data. In other cases, organizations may be interested in finding *any* feasible solution to large, vexing problems.

Note that our Acme example only specified two digits of accuracy for resource usage and may have only one digit of accuracy for profitability per product.

Note that in our earlier example, if we had a wide enough suboptimality tolerance, say 30% and followed the branch for $x_1 \geq 3.0$ first rather than $x_1 \leq 2.0$, we might have terminated with a suboptimal solution.

6.5 Binary Variables and Logical Relations

Binary variables are a special case of general integer variables. Rather than variables taking on values such as 46 and 973, acceptable values are limited to just 0 and 1. This greatly reduces the possible search space for branch and bound since you know that in any branch, you will never branch on a single variable more than once. On the other hand, with a large number of variables, the search space can still be very large. If the previous case with 1000 variables were binary, a full enumeration list would rely on a list of 2^{1000} or one followed by about 300 zeros) possible solutions. While better than the case of integers, it is still vastly more than the number of atoms in the universe.

TABLE 6.15 Data for the Outtel Example

	Inv	NPV
A	12	2.0
B	24	3.6
C	20	3.2
D	8	1.6
E	16	2.8

The important thing is that binary variables give us a lot of rich opportunities to model complex, real world situations.
Examples include assigning people to projects, go-no go decisions on projects, and much more.

Let's explore the impact of binary restrictions with another example. The world famous semiconductor company, Outtel has a choice of five major R&D projects:

1. Develop processor architecture for autonomous vehicles
2. Next generation microprocessor architecture
3. Next generation process technology
4. New fabrication facility
5. New interconnect technology

The key data is described as follows. *NPV* is Net Present Value of the project factoring in costs. *Inv* is the capital expenditures or investment required for each project. The company has $40 Billion in capital available to fund a portfolio of projects. Let's set up the data as matrices in R.

```
I <- matrix(c(12,24,20,8, 16), ncol=1,
            dimnames=list(LETTERS[1:5],"Inv"))
N <- matrix(c(2.0, 3.6, 3.2, 1.6, 2.8), ncol=1,
            dimnames=list(LETTERS[1:5],"NPV"))
kbl (cbind(I,N), booktabs=T,
    caption="Data for the Outtel Example")
```

These projects all carry their own expenses and engineering support. There are limits to both the capital investment and engineering resources available. To start, consider all of these project to be independent.

Exercise 6.1 (Create and Solve Model). Formulate and solve a basic, naive model that allows projects to repeated and partially completed. Implement and solve your model. (Hint: You don't need binary variables yet.) What is the optimal decision? Is this result surprising? Does this make sense in terms of the application?

Exercise 6.2 (No Project is Repeated). Now, let's explore one aspect of moving towards binary variables. What constraint or constraints are needed to ensure no project is done more than once, while allowing partial projects to still be funded and incur both proportionate costs and benefits. How does this solution compare to that of above?

Exercise 6.3 (No Partial Projects). What change is needed to prevent partial funding of projects. How does this solution compare to that of above? Can you envision a situation where it might make sense to have partially funded projects? Could this be similar to corporate joint ventures where investments and benefits are shared across multiple entities?

Now that you have binary constraints, let's explore the impact of logical restrictions with a series of separate scenarios. In each of these cases, make sure that you implement the relationship as a linear relationship. This means that you cannot use a function such as an "If-then", "OR", "Exclusive OR", or other relationship. Furthermore, you can't multiply variables together or divide variables into variables.

Let's assume that projects A and B require the focused effort of Outtel's top chip architects and Outtel has decided, therefore, that it is not possible to do both projects.

Therefore, they want to ensure that at most one of the two projects can be pursued. What needs to be added to enforce this situation?

Let's start with a truth table summarizing the interaction between projects (variable values) and the "Top Chip Architects" relationship.

TABLE 6.16 Relationships between Projects A, B, and the Need for Chip Architects

Project A y_A	Project B y_B	Violates "Top Chip Architects" Relationship
0	0	No
0	1	No
1	0	No
1	**1**	**Yes**

Now we need to create a constraint that blocks the situations that violate the relationship ($y_A = y_B = 1$) but allows the other situations to be unblocked. Recall that you want to have a linear constraint or constraints that would be added to the model. In this case, we can use $y_A + y_B \leq 1$. It is a good exercise to check this relationship against the truth table.

Note that we cannot use $y_A \cdot y_B = 0$ because it is a nonlinear constraint when we multiply two variables together. Sometimes trial and error is needed to verify that a constraint enforces the needed situation. In these case, it can be helpful to draw up a truth table with values for the associated variables and see whether the satisfies the requirement.

In each of the following sections create a linear relationship that models this situation.

Exercise 6.4 (Interconnects Need Something to Connect). Instead of the constraint on chip architects, let's consider the situation of the interconnect technology. Assume that project E on interconnect technology would only provide strong benefit if a new architecture is also developed. In other words, E can only be done if A or B is done. Note that if both A and B are done, E will certainly have a use!

Exercise 6.5 (Manufacturing). Outtel knows that staying aggressive in manufacturing is important in this competitive industry but that it is too risky to do too much at once from manufacturing perspective. The result is that Outtel want to ensure that exactly one major manufacturing initiative (C or D) must be done. In other words, project C or project D must be done but not both.

Exercise 6.6 (Solving Each Case). Solve for the base case and then show the results of just each constraint at once. Combine the results into a single, well-formatted table. Discuss the results.

Exercise 6.7 (Full Enumeration). When projects can be neither partially funded nor repeated, how many candidate solutions would there be by full enumeration? List them.

6.6 Fixed Charge Models

Fixed charge models are a special case of integer programming models where situations where product cost has a fixed cost component and a marginal (per unit) cost component. A common example of this is when a machine must be setup before any production can occur.

6.6.1 Fixed Charge Example-Introduction

In our example, we need to connect or *link* the two decision variables of how much to produce and whether to produce for each of the products.[2]

Let's explore an example of a fixed charge problem. Widget Inc. is re-evaluating their product production mix. As the plant manager, you are responsible for determining what products the company should manufacture. Since the company is leasing equipment, there are setup costs for each product. You need to determine the mix that will maximize profit for the company.

The profit for each Widget is shown below, as well as the setup costs (if you decide to produce any Widgets at all.) The materials you have sourced allow you to only produce each Widget up to its capacity.

TABLE 6.17 Widget Characteristics

Product	Profit	Setup Cost	Capacity
Widget 1	\$15	\$1000	500
Widget 2	\$10	\$1500	1000
Widget 3	\$25	\$2000	750

To produce each Widget, the hours required at each step of the manufacturing process are shown below. Also shown are the availability (in hours) of the equipment at each step.

TABLE 6.18 Widget Production Requirements

Production	Widget 1	Widget 2	Widget 3	Available
Pre-process	2	1	3	1000
Machining	1	5	2	2000
Assembly	2	1	1	1000
Quality Assurance	3	2	1	1500

In order to develop our optimization model, we determine the objective function (goal), decision variables (decisions) and constraints (limitations).

[2]This example is based on an example from Thanh Thuy Nguyen.

Objective Function: Maximize net profit

Decision Variables:

- W_1 = Number of Widget 1 to produce

- W_2 = Number of Widget 2 to produce
- W_3 = Number of Widget 3 to produce

- Y_1 = 1 if you choose to produce Widget 1; 0 otherwise
- Y_2 = 1 if you choose to produce Widget 2; 0 otherwise
- Y_3 = 1 if you choose to produce Widget 3; 0 otherwise

Constraints:

1. Using no more than 1000 hours of Pre-process
2. Using no more than 2000 hours of Machining
3. Using no more than 1000 hours of Assembly
4. Using no more than 1500 hours of Quality Assurance
5. Producing no more than 500 of Widget 1
6. Producing no more than 1000 of Widget 2
7. Producing no more than 750 of Widget 3

We can then write this as the following formulation.

$$\begin{aligned}
\text{Max } & 15W_1 + 10W_2 + 25W_3 - 1000Y_1 - 1500Y_2 - 2000Y_3 \\
\text{s.t.: } & 2W_1 + 1W_2 + 3W_3 \leq 1000 \\
& 1W_1 + 5W_2 + 2W_3 \leq 2000 \\
& 2W_1 + 1W_2 + 1W_3 \leq 1000 \\
& 3W_1 + 2W_2 + 1W_3 \leq 1500 \\
& W_1 \leq 500 \\
& W_2 \leq 1000 \\
& W_3 \leq 750 \\
& W_1,\ W_2,\ W_3 \geq 0 \\
& Y_1,\ Y_2,\ Y_3\ \in \{0, 1\}
\end{aligned}$$

Let's examine the results if we run this model.

For fixed charge problems, we need to link our Widget production volume to our decision on whether or not we produce the Widget (and its associated setup costs). We do this what is called the "Big M" method and the aptly named *linking constraint.*

```
fc_base_model <- MIPModel() |>
 add_variable(Vw1, type="integer", lb=0, ub=500)  |>
 add_variable(Vw2, type="integer", lb=0, ub=1000) |>
 add_variable(Vw3, type="integer", lb=0, ub=750)  |>
 add_variable(Vy1, type="binary")                 |>
    # Binary Decision for Widget 1 Setup
 add_variable(Vy2, type = "binary")               |>
    # Binary Decision for Widget 2 Setup
 add_variable(Vy3, type = "binary")               |>
    # Binary Decision for Widget 3 Setup

 set_objective(15*Vw1 + 10*Vw2 + 25*Vw3
               - 1000*Vy1 - 1500*Vy2 - 2000*Vy3,
               "max")                            |>
 add_constraint(2*Vw1 + 1*Vw2 + 3*Vw3 <= 1000) |> # Pre-process
 add_constraint(1*Vw1 + 5*Vw2 + 2*Vw3 <= 2000) |> # Machining
 add_constraint(2*Vw1 + 1*Vw2 + 1*Vw3 <= 1000) |> # Assembly
 add_constraint(3*Vw1 + 2*Vw2 + 1*Vw3 <= 1500)    # Quality

fc_base_res <- solve_model(fc_base_model,
                           with_ROI(solver = "glpk"))
fc_base_res
```

```
## Status: optimal
## Objective value: 8845
```

Our analysis found an optimal solution with a profit of 8845. At first glance, this may look good but let's examine the results in more detail.

```
fc_base_summary <-
  cbind(fc_base_res$objective_value,
        t(as.matrix(fc_base_res$solution)))
colnames(fc_base_summary)<-
  c("Net Profit", "$w_1$","$w_2$","$w_3$",
    "$y_1$","$y_2$","$y_3$")
kbl (fc_base_summary, booktabs=T, escape=F,
     caption = "Base Case Solution for Fixed Charge Problem") |>
  kable_styling(latex_options = "hold_position") |>
  footnote ("Note that no setups are incurred.")
```

We are producing a mix two products w_2 and w_3 are positive but $y_2 = y_3 = 0$. This is a nice, very high profit situation because we are not paying for

TABLE 6.19 Base Case Solution for Fixed Charge Problem

Net Profit	w_1	w_2	w_3	y_1	y_2	y_3
8845	0	307	231	0	0	0

Note:
Note that no setups are incurred.

the production setups. It may not be surprising that the optimization model chooses to avoid "paying" the fixed charge for production. This is a "penalty" in the objective function. What we need is a way of connecting the amount to produce of a widget and the decision to produce any of that widget. In reality, it would be necessary to pay the setup costs for the second and third products $(1500 + 2500)$ bringing the total profit down to 5345. Since the model is not factoring in the production setup cost, perhaps a higher net profit could be found if we could directly account this in our product offerings.

What we need is a way to connect, associate, or dare I say "link" each production decision, w, with its setup decision, y. In fact, this connection is called a *linking constraint* and is quite common in mixed integer programming problems.

Let's return to the idea of the truth table – let's focus on widget 2.

TABLE 6.20 Truth Table for Widget 2 Variables

Amount to Produce w_2	Decision to Produce y_2	Interpretation of situation
0	0	OK – choosing to do nothing
0	1	OK – perhaps not a "smart" option but not impossible or unheard of
1	0	**impossible** – can't produce even a single sellable product without doing a setup
1	1	OK
42	0	**impossible**
42	1	OK

For illustration purposes, I pick three different widget 2 production values (w_2). The value of $w_2 = 0$ means that we aren't actually producing anything. In this case, it doesn't matter whether we do a setup $(y_2 = 1)$ or not $(y_2 = 0)$. Some people might be offended by a situation of not producing anything but paying for a setup. The objective function will discourage this situation from happening so as to avoid paying for a setup, but it is possible that there might be other relationships between products. In addition, we are all probably familiar with organizations that have invested in a product, only to never put it in service. We don't need to make the constraint enforce smart operations.

To demonstrate the relationship between the decisions for widget 2, it is also helpful to look at the small volume case of producing just 1 product ($w_2 = 1$). The key situation here is to ensure that if we even make a single product, must not allow the optimization model to avoid doing a setup. Along the way, it is good to make sure that the relationship allows the solution of $w_2 = 1$ and $y_2 = 1$ to be considered feasible.

Lastly, I like to include a larger production volume to ensure that the constraint is properly enforcing this relationship. The value, 42, is just an arbitrary one and could be any other moderately large value.

6.6.2 Linking Constraints with "Big M"

The linking constraint as the following: $w_1 \leq M \cdot y_1$ The value M is a big value that is so large that it does not prematurely limit the production for widget 1. Whatever value of M is used, w_1 can never exceed that value.

Our linking constraints force our new values to be 0 or 1. If $w_1 > 0$, then this constraint forces the associated y_1 to be equal to 1. If $w_1 = 0$, then this constraint allows y_1 to be either 0 or 1. However, our objective function will cause Solver to avoid paying a setup by setting $y_1 = 0$.

It might be tempting to select a very large number such as a billion 10^{10} but picking excessively large numbers, can result in poor computational performance. As we've discussed large, real-world optimization problems are hard enough, let's not make it any harder. Albert Einstein once said "A model should be as simple as possible, but no simpler." Essentially, a value for M should be as small as possible, but no smaller.

Since M serves to impose an upper bound on w_1, this might suggest how we can use this information to pick an appropriate value for M and that we can do so for each product separately, using a separate value M_i for each widget.

Let's examine the constraints for the production plan in the situation gives all resources to the production of Widget 1 and nothing to Widget 2 and Widget 3.

More formally, using the non-negative lower bounds on widget production, we can set $w_2 = w_3 = 0$ and substitute into the constraints of the optimization model. The constraints now look like the following:

$$\begin{aligned}
2 \cdot w_1 + 1 \cdot 0 + 3 \cdot 0 &\leq 1000 \\
1 \cdot w_1 + 5 \cdot 0 + 2 \cdot 0 &\leq 2000 \\
2 \cdot w_1 + 1 \cdot 0 + 1 \cdot 0 &\leq 1000 \\
3 \cdot w_1 + 2 \cdot 0 + 1 \cdot 0 &\leq 1500 \\
w_1 &\leq 500 \\
0 &\leq 1000 \\
0 &\leq 750
\end{aligned}$$

This can be further simplified as the following.

$$\begin{aligned}
2 \cdot w_1 &\leq 1000 \\
1 \cdot w_1 &\leq 2000 \\
2 \cdot w_1 &\leq 1000 \\
3 \cdot w_1 &\leq 1500 \\
w_1 &\leq 500
\end{aligned}$$

It can be further simplified as the following:

$$\begin{aligned}
w_1 &\leq \frac{1000}{2} = 500 \\
w_1 &\leq 2000 \\
w_1 &\leq \frac{1000}{2} = 500 \\
w_1 &\leq \frac{1500}{3} = 500 \\
w_1 &\leq 500
\end{aligned}$$

Since all constraints need to be satisfied, we can represent this to mean that Widget 1 manufacturing could never be higher than the smallest (most restrictive) of these constraints.

Therefore,

$$w_1 \leq min\{\frac{1000}{2}, \frac{2000}{1}, \frac{1000}{2}, \frac{1500}{3}, 500\}$$

or

$$w_1 \leq min\{500, 2000, 500, 500, 500\} = 500$$

Therefore, we know that Widget 1 production can never exceed 500 units regardless of the amount of other Widgets (w_2 and w_3) produced. We can then safely set the Big M value for Widget 1 to be 500. In other words,

$M_1 = 500$.

We can follow the same process for setting Big M values for widget 2. We start by setting $w_1 = w_3 = 0$. We know that w_2 must be no larger than the most restrict constraint. I'll skip rewriting the constraints and jump a little ahead.

$$W_2 \leq min\{\frac{1000}{1}, \frac{2000}{5}, \frac{1000}{1}, \frac{1500}{2}, 1000\} = 400$$

Therefore, we can safely use a Big M value for w_2 of $M_2 = 400$.

Again we can follow the same process for w_3 to find a small Big M value.

$$w_3 \leq min\{\frac{1000}{3}, \frac{2000}{2}, \frac{1000}{1}, \frac{1500}{1}, 750\} = 333.33$$

Since the Big M value is setting a lower bound, in this way, we can round up from 333.33 to 334. Our updated model with "Big M":

$$\begin{aligned}
&\text{Max } 15w_1 + 10w_2 + 25w_3 - 1000y_1 - 1500y_2 - 2000y_3 \\
&\text{s.t.: } 2w_1 + 1w_2 + 3w_3 \leq 1000 \\
&\quad 1w_1 + 5w_2 + 2w_3 \leq 2000 \\
&\quad 2w_1 + 1w_2 + 1w_3 \leq 1000 \\
&\quad 3w_1 + 2w_2 + 1w_3 \leq 1500 \\
&\quad w_1 \leq 500 \\
&\quad w_2 \leq 1000 \\
&\quad w_3 \leq 750 \\
&\quad w_1 - 500y_1 \leq 0 \\
&\quad w_2 - 400y_2 \leq 0 \\
&\quad w_3 - 334y_3 \leq 0 \\
&\quad w_1, w_2, w_3 \geq 0 \\
&\quad y_1, y_2, y_3 \in \{0, 1\}
\end{aligned}$$

We could optionally delete the production capacity such as $w_1 \leq 500$ as they are now redundant constraints because the fixed charge constraints also reflect these upper limits on production capacity.

6.6.3 Fixed Charge Implementation

Given that we have already created a model without the Big M constraints, we can simply add the constraints to the model. We'll skip the piping operator and just add the constraint directly to the previous model, `fc_base_model`.

```
fc_bigM_model <- add_constraint(fc_base_model,
                               Vw1 - 500*Vy1 <= 0) |>
           # W1's Big M linking constraint
 add_constraint(Vw2 - 400*Vy2 <= 0)                |>
           # W2's Big M linking constraint
 add_constraint(Vw3 - 334*Vy3 <= 0)
           # W3's Big M linking constraint

fc_bigM_res <- solve_model(fc_bigM_model,
                           with_ROI(solver = "glpk"))
fc_bigM_res
```

```
## Status: optimal
## Objective value: 6500
```

Our model was able to find an optimal solution with an objective value of 6500. Our optimal production plan is shown below. As we can see, our model returned a production plan with only one model of widget being produced.

Interesting, the objective function value has gone down significantly. Let's look this over in more detail and compare it to the results that we had when we avoided paying for setups.

TABLE 6.21 Fixed Charge Problem

	Net Profit	w_1	w_2	w_3	y_1	y_2	y_3
Base Case w/o Setup	8845	0	307	231	0	0	0
Base Case with Setup	5345	0	307	231	0	0	0
Optimal Fixed Charge	6500	500	0	0	1	0	0

Notice that the high setup costs have caused us to focus our production planning decisions. Rather than spreading ourselves across three different product lines, we are producing as many of widget 1 as we can.

6.7 Model Results and Interpretation

We can also calculate our resource usage.

```
fc_bigM_res.W1 <- get_solution(fc_bigM_res, Vw1)
        # Extract solution value for decision variable, W1
fc_bigM_res.W2 <- get_solution(fc_bigM_res, Vw2)
        # Extract solution value for decision variable, W2
fc_bigM_res.W3 <- get_solution(fc_bigM_res, Vw3)
        # Extract solution value for decision variable, W3
fc_bigM_res.df <- data.frame(c(fc_bigM_res.W1,
                              fc_bigM_res.W2,
                              fc_bigM_res.W3))
fc_bigM_res.r1 <- t(data.frame(c(fc_bigM_res.W1*2,
                                fc_bigM_res.W2*1,
                                fc_bigM_res.W3*3)))
               #multiply results with hours used
rownames(fc_bigM_res.r1) <- "Pre-process"
fc_bigM_res.r2 <- t(data.frame(c(fc_bigM_res.W1*1,
                                fc_bigM_res.W2*5,
                                fc_bigM_res.W3*2)))
               #multiply results with hours used
rownames(fc_bigM_res.r2) <- "Machining"
fc_bigM_res.r3 <- t(data.frame(c(fc_bigM_res.W1*2,
                                fc_bigM_res.W2*1,
                                fc_bigM_res.W3*1)))
               #multiply results with hours used
rownames(fc_bigM_res.r3) <- "Assembly"
fc_bigM_res.r4 <- t(data.frame(c(fc_bigM_res.W1*3,
                                fc_bigM_res.W2*2,
                                fc_bigM_res.W3*1)))
               #multiply results with hours used
rownames(fc_bigM_res.r4) <- "Quality Assurance"
fc_bigM_res.tot <- data.frame(c(sum(fc_bigM_res.r1),
                               sum(fc_bigM_res.r2),
                               sum(fc_bigM_res.r3),
                               sum(fc_bigM_res.r4)))
               #sum each step
colnames(fc_bigM_res.tot) <- "Total Used"
fc_bigM_res.avail <- data.frame(
  c("1000","2000","1000","1500"))
```

```
          #print available hours for each step
colnames(fc_bigM_res.avail) <- "Available"

Res_Usage <- cbind(fc_bigM_res.tot, fc_bigM_res.avail)
rownames(Res_Usage)<-c("Pre-process", "Machining",
                       "Assembly", "Quality Assurance")
kbl (cbind(Res_Usage),
     booktabs=T, caption =
       "Manufacturing Resource Usage") |>
  kable_styling(latex_options = "hold_position")
```

TABLE 6.22 Manufacturing Resource Usage

	Total Used	Available
Pre-process	1000	1000
Machining	500	2000
Assembly	1000	1000
Quality Assurance	1500	1500

As can be seen, we used all of our Pre-processing hours, Assembly hours, and Quality Assurance hours. There is a significant amount of time available in Machining though.

Further analysis could examine alternatives such as redesigning widget 2 and 3 to be less resource intensive in production to see at what point we would choose to produce them.

Exercise 6.8 (Redesign Widget 3)**.** The design team has an idea of how to use a previous manufacturing fixture which could eliminate the setup cost for Widget 3. How much would you produce of Widget 3 if there were no setup (fixed) cost for production of Widget 3? Of course this still keeps a cost for Widgets 1 and 2. Experiment with the above model to find the lowest setup cost for Widget 3 where you still choose to not produce any of Widget 3.

Exercise 6.9 (Reducing Fixed Costs)**.** How would production change if the setup costs were cut in half due to implementing lean production approaches?

Exercise 6.10 (Adding a Fourth Widget)**.** Consider adding a new widget 4 which needs 4, 3, 2, and 5 hours for Pre-processing, Machining, Assembly, and Quality Assurance, respectively. Widget 4 uses the same setup as Widget 3 which means no repeated setup cost for Widget 4 if widget 3 is already being produced.

Solve the model to obtain the production plan where both Widget 3 and Widget 4 are produced.

7

More Integer Programming Models

7.1 Overview

This chapter consists of a collection of rich MILP models that can be used for inspiration on products. Some of these cases are expansions of Dirk Schumacher's `ompr`vignettes. These make for excellent resources demonstrating a variety of features such as creation of simulated data and visualization of results. I strongly recommend reading the original. These vignettes can be viewed from the package documentation, Dirk Schumacher's github web site[1], or downloading his github repository.

- All text from Dirk Schumacher's articles are set in block quotes.
- Code chunks are generally based on Dirk Schumacher's articles unless stated otherwise.
- The LaTeX formulations are generally based on Dirk Schumacher's LaTeX with some modifications to support LaTeX environments.

7.2 Revisiting the Warehouse Location Problem

Let's start by reviewing Dirk's original description of the warehouse location problem.

> In this article we will look at the Warehouse Location Problem[2]. Given a set of customers and set of locations to build warehouses, the task is to decide where to build warehouses and from what warehouses goods should be shipped to which customer.

[1] `https://dirkschumacher.github.io/ompr/articles/index.html`
[2] `https://en.wikipedia.org/wiki/Facility_location_problem`

DOI: 10.1201/9781003051251-7

Thus there are two decisions that need to made at once: where and if to build warehouses and the assignment of customers to warehouses. This simple setting also implies that at least one warehouse must be built and that any warehouse is big enough to serve all customers.

As a practical example: you run the logistics for an NGO and want to regularly distribute goods to people in need. You identified a set of possible locations to set up your distribution hubs, but you are not sure where to build them. Then such a model might help. In practice however you might need to incorporate additional constraints into the model. Let's start by defining the decision variables. Each possible location of a warehouse, j can have a warehouse be built or not be built. We will use $y_j = 1$ to indicate that warehouse j is built. Conversely, $y_j = 0$ indicates that we are not building a warehouse at location j. Since a warehouse is either built or not built, a binary variable is appropriate for y_j.

Similarly, the variable $x_{i,j}$ is the decision of assigning customer i to warehouse j. It also needs to be a binary variable since partial assignments are not allowed. Therefore, $x_{3,7} = 1$ means that customer *3* is assigned to warehouse *7*, and we would expect $x_{3,8} = 0$ since a customer can't be assigned to two warehouses.

Now, that we have a handle on the variables, let's move on to the constraints. Each customer must be assigned to one and only one warehouse. For illustration, this means that one of the variables $x_{1,j}$ must be set to one and the others are zero. To enforce this constraint, we can simply add the x variables for warehouse 1 for each of the warehouses. We could do this with $x_{1,1} + x_{1,2} + ... + x_{1,m}$ and requiring it to be one. We could rewrite this using a summation as $\sum\limits_{j=1}^{m} x_{1,j} = 1$. That constraint is limited to just customer 1 though. We don't want to write this constraint out separately for each customer so we can generalize this by repeating it for all n customers as $\sum\limits_{j=1}^{m} x_{i,j} = 1,\ i = 1, ..., n.$

It would not work to assign a customer to an unbuilt warehouse. For example, it would not work to have customer 23 assigned to warehouse 7 if warehouse 7 was not built. In other words, the combination of $x_{23,7} = 1$ and $y_7 = 0$ would be a big problem and should not happen. We need to connect the decision variable $x_{23,7}$ and y_7. One way to do that is creating a constraint, $x_{23,7} \leq y_7$ which explicitly blocks assigning customer 23 to warehouse 7 unless warehouse 7 is operating. This can be generalized as $x_{i,j} \leq y_j,\ i = 1, \ldots, n,\ j = 1, \ldots, m$.

Our objective is to minimize cost. We have a cost assumed for each customer to warehouse assignment. This might be related to distance. Let's refer to this cost as $\text{transportcost}_{i,j}$. The cost based on the transportation from warehouse to customer is then $\sum_{i=1}^{n} \sum_{j=1}^{m} \text{transportcost}_{i,j} \cdot x_{i,j}$. Note that we could concisely abbreviate it as something like $C^T_{i,j}$. In this case, using a capital C to indicate cost and a superscript T to indicate transportation cost.

We also have a cost factor for each warehouse that we choose to build/operate. Again, if we define fixedcost_j as the fixed cost for building warehouse j, the cost of our decisions is simply, $\sum_{j=1}^{m} \text{fixedcost}_j \cdot y_j$.

Let's combine all this together into the formulation.

$$
\begin{aligned}
\min \ & \sum_{i=1}^{n} \sum_{j=1}^{m} \text{transportcost}_{i,j} \cdot x_{i,j} + \sum_{j=1}^{m} \text{fixedcost}_j \cdot y_j \\
\text{s.t.} \ & \sum_{j=1}^{m} x_{i,j} = 1,\ i = 1, \ldots, n \\
& x_{i,j} \leq y_j,\ i = 1, \ldots, n,\ j = 1, \ldots, m \\
& x_{i,j} \in \{0, 1\}\ i = 1, \ldots, n,\ j = 1, \ldots, m \\
& y_j \in \{0, 1\}\ j = 1, \ldots, m
\end{aligned}
$$

7.2.1 Implementing the Warehouse Model

Rather than typing in fixed data or loading it from data file, Dirk simply generated a collection of random data for testing purposes. This highlights the advantage of R in that we have a rich collection of numerical tools available that we can leverage.

The first thing we need is the data. In this article we will simply generate artificial data.

We assume the **customers** are located in a grid with Euclidian

distances. Let's explain the functions used for this data generation. The first command sets the random number seed. In general, computers don't actually generate random numbers, they generate what is called pseudo random numbers according to a particular algorithm in a sequence. The starting point will set a sequence that appears to be random and behaves in a relatively random way. Much more can be said but this is setting as the first random number, 1234. It could just as easily have been 4321 or any other integer.
Let's unpack this further.

The grid size is set to 1000. You can think of this as a map for a square region that is 1000 kilometers in both horizontal (East-West) and vertical (North-South) directions for visualization purposes. We are then randomly placing customers on the map.

Next, we set the number of customers to be 200 and the number of warehouses to be 40. The size of this problem can have a tremendous impact on solving speed.[3]

```
set.seed(1234)        # Set random number seed
grid_size <- 1000     # Horizontal and vertical axis ranges
n <- 200 ; m <- 40    # Set number of customers and warehouses
```

Next, we create the customer data as a dataframe. The `runif` function generates a uniform random number between 0.0 and 1.0 such as perhaps 0.32721.This would be multiplied by our grid size multiplier of 1000 resulting in 327.21. This is then rounded to 327. Let's now see how this is implemented for customers.

```
customer_locations <- data.frame(
  id = 1:n,                         # Create column with ID numbers from 1 to n
  x = round(runif(n) * grid_size), # Create n x-coordinate values
  y = round(runif(n) * grid_size)  # Create n y-coordinate values
)
```

[3]In particular, I recommend using a much smaller problem size (fewer customers and warehouses) if you are using a slow computer or a shared resource such as a server or cloud-based service like RStudio.cloud. Integer programming problems can quickly consume tremendous amounts of computer processing time which can have budgetary impacts or other problems.

Choosing to open a warehouse entails a fixed cost that varies for each warehouse, j, defined earlier as $fixedcost_j$ or `fixedcost[j]` in `R`. The warehouse locations and fixed costs are again generated randomly as described by Dirk Schumacher.

The **warehouses** are also randomly placed on the grid. The fixed cost for the warehouses are randomly generated as well with mean cost of 10,000. Notice that in this case, the fixedcost is generated using a normal random variable function, with a mean of 10 times the grid size or 10,000 and a standard deviation of 5000.

```
warehouse_locations <- data.frame(
  id = 1:m,
  x = round(runif(m) * grid_size),
  y = round(runif(m) * grid_size)
)
fixedcost <- round(rnorm(m, mean = grid_size * 10,
                         sd = grid_size * 5))
```

The fixed costs to set up a warehouse are the following:

```
hist(fixedcost, main="Fixed Costs of Warehouses",
     xlab="Cost")
```

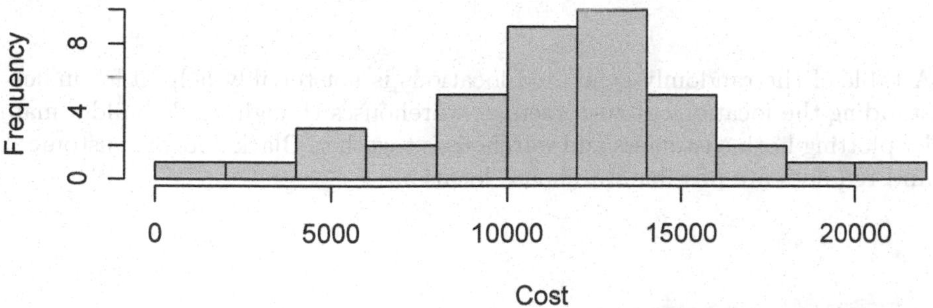

he distribution of fixed costs is pretty broad. In fact, using a normal distribution means that is unbounded in both directions, so it is possible for a warehouse to have a negative construction cost. While a negative construction cost might seem impossible, perhaps this would be an example of local tax incentives at work.

For transportation cost, we will assume that simple Euclidean distance is a good measure of cost. Just a reminder, Euclidean distance between two points is $\sqrt{(X_1 - X_2)^2 + (Y_1 - Y_2)^2}$. Of course in terms of cost, we could scale it or add a startup cost to each trip but for the sake of this example, we don't need to worry about it. We could also use other distance metrics such rectilinear distance, driving distance, or any other metric we want.

Here we define a function to calculate the distance between customer i and warehouse j.

```
transportcost <- function(i, j) {
  customer <- customer_locations[i, ]
  warehouse <- warehouse_locations[j, ]
  round(sqrt((customer$x - warehouse$x)^2 +
               (customer$y - warehouse$y)^2))
}
transportcost(1, 3)
```

```
## [1] 648
```

We could look at a sample of the customer location data.

```
kbl (head(customer_locations), booktabs=T,
      caption="Locations of First Six Customers") |>
  kable_styling(latex_options = "hold_position")
```

A table of the randomly generated locations is not terribly helpful for understanding the locations of customers or warehouses though. Let's build a map by plotting both customers and warehouses together. Black dots are customers and red dots are possible warehouse locations.

TABLE 7.1 Locations of First Six Customers

Id	x	y
1	114	661
2	622	528
3	609	317
4	623	768
5	861	526
6	640	732

```
p <- ggplot(customer_locations, aes(x, y)) +
  geom_point() +
  geom_point(data = warehouse_locations, color = "red",
             alpha = 0.5, shape = 17) +
  scale_x_continuous(limits = c(0, grid_size)) +
  scale_y_continuous(limits = c(0, grid_size)) +
  labs(caption="Black dots are customers. Light red triangles
       show potential warehouse locations.") +
  theme(axis.title = element_blank(),
        axis.ticks = element_blank(),
        axis.text = element_blank(),
        panel.grid = element_blank())
p
```

Note that I have modified the variables names for x and y to be `Vx` and `Vy` in the code chunk to remove confusion over x and y representing location or variables. The prefix of `V` is meant to suggest that it is a variable. It also helps prevent name space conflicts with other `R` objects in my environment.

```
model <- MIPModel() |>
  # 1 iff i gets assigned to warehouse j
  add_variable(Vx [i, j], i=1:n, j=1:m, type="binary") |>

  # 1 iff warehouse j is built
  add_variable(Vy [j], j = 1:m, type = "binary")        |>

  # minimize cost
  set_objective(sum_expr(transportcost(i, j) * Vx[i, j],
                        i = 1:n, j = 1:m) +
```

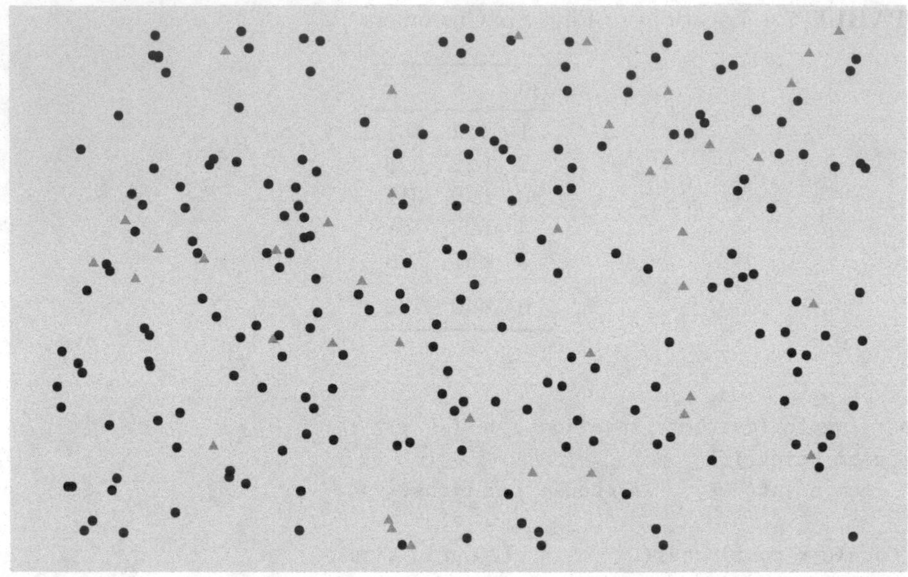

FIGURE 7.1 Map of Customer and Warehouse Locations.

```
                  sum_expr(fixedcost[j] * Vy[j], j = 1:m),
                "min")                                    |>

  # every customer needs to be assigned to a warehouse
  add_constraint(sum_expr(Vx[i, j], j=1:m)==1, i=1:n) |>

  # if a customer is assigned to a warehouse,
  #   then this warehouse must be built
  add_constraint(Vx[i,j] <= Vy[j], i = 1:n, j = 1:m)

model
```

```
## Mixed integer linear optimization problem
## Variables:
##   Continuous: 0
##   Integer: 0
##   Binary: 8040
## Model sense: minimize
## Constraints: 8200
```

The number of x (or `Vx`) variables is $m \cdot n = 40 \cdot 200$ or 8000 and the number of y (or `Vy`) variables is m or 20, the total number of variables is 8040 which matches the model summary. A similar calculation can be done on the constraints. The number of variables and constraints make this a non-trivial sized MIP problem. Fortunately, solving still turns out to be pretty easy on a personal computer but be cautious of solving this on a cloud instance. In fact, preparing the LP to be solved using `ompr` can take a significant amount of time and this is an area of `ompr` being worked upon.

7.2.2 Solving the Warehouse Location Problem

Now that have prepared the model, we are now ready to work on solving the model. We will start with using `glpk` as we have done in previous examples.

The solver can generate a lot of additional information. We previously kept the default of `verbose = FALSE` but let's see what we get with a more talkative or verbose option set (`verbose = TRUE`).

So far, you are unlikely to have noticed an optimization problem cause a visible delay. If you run the following code chunk to solve this model, you may see the first distinct pause. As I said, this is a non-trivial MIP problem! It may take only a few seconds or perhaps a few minutes depending upon your computer. This also means that you must be careful in running MIP problems on cloud based services. A particularly big MIP running on Amazon Web Services might create a sizable bill. This is the case whether you are using an Amazon Web Services instance, RStudio.cloud, or another cloud/shared service.

In summary, non-trivial MIP models should be run locally on a computer unless you are certain that your computational loads will not cause a problem for yourself or others.

```
result <- solve_model(model, with_ROI(solver = "glpk"))

result
```

```
## Status: optimal
## Objective value: 60736
```

We can see that the result was solved to optimality and the objective function value of We can summarize the information by simply extracting a the objective function value. This can be done with inline R code. For example, we can say the following.

We solved the problem with an objective value of 60736.

With 8040 variables, it isn't always terribly useful to just list out all of the variables.

At this point, we are really interested in the small minority of non-zero variables. Only a small percentage of the x variables are not zero.[4]

We can do some processing to extract the non-zero x (or `Vx`) variables from those with zero values. In order to do so, Dirk uses the `dplyr` package. This package provides a tremendous number of functions for data management. To illustrate how this is done, let's review line by line how this code chunk works.

```
matching <- result |>
  get_solution(Vx[i,j]) |>
  filter(value > .9) |>
  select(i, j)
```

The next command `matching <- result |>` does a few things. It uses the piping operator `|>` to pass the `result` object into being used by the following command and then everything that gets done later in this piped sequence will be assigned to `matching`.

The command `get_solution(Vx[i,j]) |>` extracts solution values for the x variables (`Vx`) from the solved model object `result` piped in from the previous stage. This will give us a data object consisting of 2000 variables values, 95% of them being zero.

The command `filter(value > .9) |>` takes the previous command's 2000 variable values and only keeps the ones for which the value is larger than 0.9. Since these are binary variables, they should all be exactly 1.0 but just in case there is any numerical anomaly where a value is passed as a value close to one, such as 0.9999999, it is best to test with some tolerance. This command then only passes to the next command the 100 non-zero values.

[4]**Challenge**: Do you know why about approximately $\frac{1}{m}$ of the decision variables are non-zero?

The last command in this code chunk `select(i, j)` says to only select (or retain) the columns named `i` and `j`. Notice that this does not pass the actual value of zero or one – we know it is one! Also, note that it does not include the piping operator `|>` which means that it is the end of this piped sequence.

Let's review what the two versions of the results look like.

```
kbl (head( get_solution(result, Vx[i,j])), booktabs=T,
      caption = "Raw Results for Vx Variable.") |>
  kable_styling(latex_options = "hold_position")
```

TABLE 7.2 Raw Results for Vx Variable

variable	i	j	value
Vx	1	1	0
Vx	2	1	0
Vx	3	1	0
Vx	4	1	0
Vx	5	1	0
Vx	6	1	0

Notice that in the table of raw results, the first six entries (obtained using the `head` function) did not show any customers assigned to a warehouse. This isn't surprising given that only one in twenty values are non-zero. It is common in many binary programming cases for the vast majority of variables to have a value of zero.

```
kbl (head( matching), booktabs=T,
      caption=
   "Subset of Customers (i) Assigned to Warehouses (j)") |>
  kable_styling(latex_options = "hold_position")
```

The processed results table shows that the `matching` object simply lists six customers and to which warehouse it is assigned. It has cut out the vast majority of rows (decision variables) with zero values.

This clean and simple listing from `matching` is interesting and will then be used for adding lines to the earlier `ggplot` model.

TABLE 7.3 Subset of Customers (i) Assigned to Warehouses (j)

i	j
4	3
5	3
6	3
10	3
11	3
14	3

The last step is to add the assignments to the previous plot we generated. Dirk used one single code chunk to do more processing of results, focused on warehouses, and then create the final plot. I'll break up the steps into separate code chunks just for the sake of illustrating how they work and showing the intermediate products. This is a useful chance to see how to manage results.

```
plot_assignment <- matching |>
  inner_join(customer_locations, by = c("i" = "id")) |>
  inner_join(warehouse_locations, by = c("j" = "id"))
kbl (head(plot_assignment), booktabs=T, caption=
      "XY coordinates to draw Customer-Warehouse Routes") |>
  kable_styling(latex_options = "hold_position")
```

TABLE 7.4 XY Coordinates to Draw Customer-Warehouse Routes

i	j	x.x	y.x	x.y	y.y
4	3	623	768	757	745
5	3	861	526	757	745
6	3	640	732	757	745
10	3	514	986	757	745
11	3	694	566	757	745
14	3	923	758	757	745

Notice that this gives the beginning and ending coordinates of what will soon be lines to show the connections between customers and warehouses.

Now, let's calculate how many customers each warehouse serves.

```
customer_count <- matching |>
  # Process matching and save result to customer_count
  group_by(j)                |> # Group by warehouse
  summarise(n = n())         |> # Summarize count
  rename(id = j)                # Rename column from j to id
kbl (customer_count, booktabs=T, caption=
  "Count of Customers Assigned to Each Operating Warehouse") |>
  kable_styling(latex_options = "hold_position")
```

TABLE 7.5 Count of Customers Assigned to Each Operating Warehouse

id	n
3	67
24	61
28	72

We can see that warehouses selected to be built serve different numbers of customers. Let's gather more information to examine the fixed cost of the selected. How does this compare with the histogram of fixed costs?

```
plot_warehouses <- warehouse_locations                     |>
  mutate(costs = fixedcost)                                |>
  inner_join(customer_count, by = "id")                    |>
  filter(id %in% unique(matching$j))
kbl (plot_warehouses, booktabs=T, caption =
        "Summary of Results for Warehouses Used") |>
  kable_styling(latex_options = "hold_position")
```

TABLE 7.6 Summary of Results for Warehouses Used

id	x	y	costs	n
3	757	745	4511	67
24	105	517	1184	61
28	426	393	4061	72

The `plot_warehouses` data frame[5] is built using `dplyr` functions to create a

[5]Note that the data type can sometimes be a source of problems for R users. You can use the command `class(plot_warehouses)` to confirm that it is a `data.frame`.

summary of the important information about the warehouses used: where they are located, costs, and number of customers served.

```
p +
  geom_segment(data=plot_assignment,
               aes(x=x.y, y=y.y, xend=x.x, yend=y.x)) +
  geom_point(data  = plot_warehouses,
             color = "red", size=3, shape=17) +
  ggrepel::geom_label_repel(data  = plot_warehouses,
                            aes(label = paste0(
                              "fixed costs:",costs, ";
                              customers: ", n)),
                            size = 2, nudge_y = 20) +
  ggtitle(paste0(
  "Optimal Warehouse Locations and Customer Assignment"),
"Big red triangles show warehouses that will be built.
Light red are unused warehouse locations.
Lines connect customers to their assigned warehouses.")
```

Optimal Warehouse Locations and Customer Assignment

Big red triangles show warehouses that will be built.
Light red are unused warehouse locations.
Lines connect customers to their assigned warehouses.

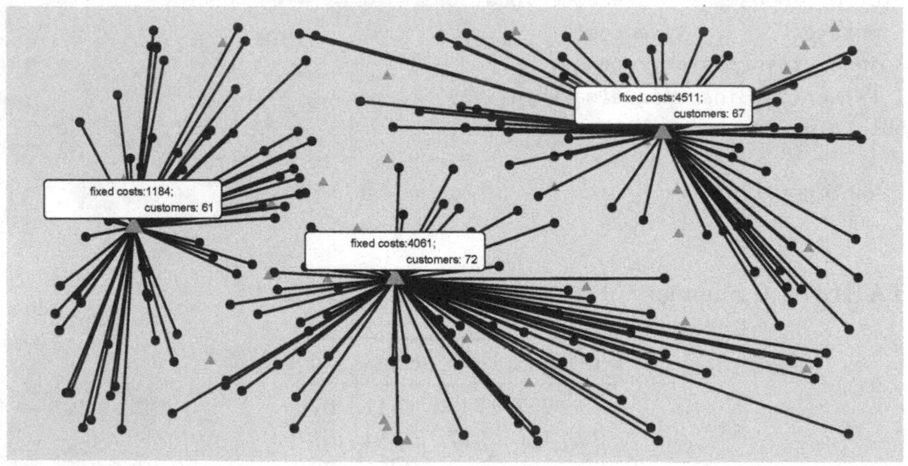

Black dots are customers. Light red triangles show potential warehouse locations.

FIGURE 7.2 Optional Warehouse Locations Customer Assignments.

The total fixed costs for setting up the 3 warehouses is:

```
sum(fixedcost[unique(matching$j)])
```

```
## [1] 9756
```

The above function cleverly uses the warehouse ID column from `matching` (`matching$j`) to make list unique and get rid of duplicates. Recall that warehouses will be listed 100 times since every customer has a warehouse listed in matching. Next, it uses these ID numbers to extract fixed-cost values and adds them up. Note that when you see a command such as this in R, it can often work to run them one at time in the console to see how they work. To demonstrate this, see how each statement builds upon the previous.

```
matching$j
```

```
##   [1]  3  3  3  3  3  3  3  3  3  3  3  3  3  3  3  3  3  3  3  3  3  3  3  3  3
##  [26]  3  3  3  3  3  3  3  3  3  3  3  3  3  3  3  3  3  3  3  3  3  3  3  3  3
##  [51]  3  3  3  3  3  3  3  3  3  3  3  3  3  3  3  3  3 24 24 24 24 24 24 24 24
##  [76] 24 24 24 24 24 24 24 24 24 24 24 24 24 24 24 24 24 24 24 24 24 24 24 24 24
## [101] 24 24 24 24 24 24 24 24 24 24 24 24 24 24 24 24 24 24 24 24 24 24 24 24 24
## [126] 24 24 24 28 28 28 28 28 28 28 28 28 28 28 28 28 28 28 28 28 28 28 28 28 28
## [151] 28 28 28 28 28 28 28 28 28 28 28 28 28 28 28 28 28 28 28 28 28 28 28 28 28
## [176] 28 28 28 28 28 28 28 28 28 28 28 28 28 28 28 28 28 28 28 28 28 28 28 28 28
```

FIGURE 7.3 LP Region and 10 IP Feasible Solutions.

```
unique(matching$j)
```

```
## [1]  3 24 28
```

```
  # Eliminate duplicate warehouse IDs
fixedcost[unique(matching$j)]
```

```
## [1] 4511 1184 4061
```

```
  #Find fixed costs of warehouses used
sum(fixedcost[unique(matching$j)])
```

```
## [1] 9756
```

```
# Add fixed costs of all warehouses used
```

7.2.3 Warehouse Discussion

This warehouse customer assignment problem was discussed in detail for multiple reasons.

- It demonstrates a large, classic, and common industrial application of optimization.
- The application uses simulated data to feed into an optimization model.
- This model uses both single and double subscripted variables as well techniques for coordinating them.
- Introduces the use of `ggplot` to visualize the data and results.
- Demonstrates data munging which is a major time sink for analysts, which can also occur in large optimizaton models so this application shows how to use `dplyr` to process data and results.

This analysis could be extended in a variety of ways to explore other questions.

- What is the statistical distribution of the optimal number of warehouses used when simulated a 1000 times?
- How often would it be optimal to five or more warehouses?
- Does decreasing variation in fixed costs affect the number of warehouses used?
- Does decreasing the mean cost of warehouses affect the number of warehouses used?

The model can also be extended in a variety of ways.

- To allow for additional application characteristics such as warehouse capacity and customer demand.
- Using actual cities in a region for locations rather than randomly generated data and a map.

The same concepts can also be applied to other applications.

7.3 Solving MIPs with Different Solvers

One of the big benefits of `ompr` is that it separates the process of formulating the optimization model from that of solving it and thereby let's us easily switch to other solvers. Up until this point, we have used the `glpk` solver. Now that we have a nontrivial optimization model, the Warehouse Location Model, let's use it.

The emphasis in this section is not model building but comparing the use of different solvers. Since we have the model already defined as an object, simply named `model`, we can easily pass it to the different solvers for optimizing. There is a large collection of other solvers that can be used with R.[6]

We'll demonstrate the performance for illustration purposes. Again, do not run the following code chunks in this section on a cloud-based service such as Rstudio.cloud, unless you are certain that imposing a non-trivial computational load is acceptable.

7.3.1 Performance of glpk

We have been using `glpk` for all of our earlier examples. For the sake of comparison, let's start off with it too. We'll set the `verbose=TRUE` option to be passed through ROI to the `glpk` solver. The exact options available for each solver and their available options varies. We will use the `tictoc` package to help us collect data on timing. Note that timing can vary even if the same problem is solved on the same computer twice in rapid succession.

```
library(tictoc)   # Package used for timing R functions
tic("glpk")        # Start the timer...
result_glpk <- solve_model(
  model, with_ROI(solver = "glpk", verbose=TRUE))
```

```
## <SOLVER MSG>  ----
## GLPK Simplex Optimizer, v4.47
## 8200 rows, 8040 columns, 24000 non-zeros
##       0: obj =  0.000000000e+000  infeas = 2.000e+002 (200)
## *   201: obj =  1.087950000e+005  infeas = 0.000e+000 (200)
## *   500: obj =  1.087950000e+005  infeas = 0.000e+000 (102)
## *  1000: obj =  1.087950000e+005  infeas = 0.000e+000 (0)
```

[6]This topic is not covered in Dirk's vignette or blog post.

```
## *  1500: obj =  1.087950000e+005  infeas = 0.000e+000 (0)
## *  2000: obj =  1.087950000e+005  infeas = 0.000e+000 (0)
## *  2500: obj =  6.119050000e+004  infeas = 0.000e+000 (0)
## *  2714: obj =  6.073600000e+004  infeas = 0.000e+000 (0)
## OPTIMAL SOLUTION FOUND
## GLPK Integer Optimizer, v4.47
## 8200 rows, 8040 columns, 24000 non-zeros
## 8040 integer variables, all of which are binary
## Integer optimization begins...
## +  2714: mip =     not found yet >=              -inf        (1; 0)
## +  2714: >>>>>  6.073600000e+004 >=  6.073600000e+004   0.0% (1; 0)
## +  2714: mip =  6.073600000e+004 >=     tree is empty   0.0% (0; 1)
## INTEGER OPTIMAL SOLUTION FOUND
## <!SOLVER MSG> ----
```

```
glpktime <- toc() # End the timer and save results
```

```
## glpk: 31.8 sec elapsed
```

```
print(solver_status(result_glpk))
```

```
## [1] "optimal"
```

```
glpktime1 <- c("glpk", glpktime$toc - glpktime$tic,
                result_glpk$status,
                result_glpk$objective_value)
```

The more verbose output provides additional details on the solving. For example, it indicates that the Simplex method was used by indicating a Simplex optimizer was used before then passing the LP relaxation solution to the integer optimization. In general, the `glpk` solver worked. We'll take the results and combine them with other shortly.

7.3.2 Performance of symphony

Let's move on to testing the `symphony` solver. Symphony is a solver with a long history of development.

```
tic("symphony")
result_symphony <-  solve_model(
  model, with_ROI(solver = "symphony", verbosity = -1))
symphonytime <- toc()
```

```
## symphony: 32.33 sec elapsed
```

```
print(solver_status(result_symphony))
```

```
## [1] "optimal"
```

```
symphonytime1 <- c("symphony",
                   symphonytime$toc - symphonytime$tic,
                   result_symphony$status,
                   result_symphony$objective_value)
```

Again, `symphony` successfully solved the optimization problem.

Several items should be noted from the above code and output.

One item is that parameters can be passed directly to solvers. This is why `verbose = TRUE` was used for `glpk` but `symphony` uses `verbosity = -1`. Differing levels of verbosity gives much more granularity than a simple TRUE/FALSE. Setting `verbosity = 0` will give rich detail that an analyst trying to improve solution speed may find useful or for debugging why a model did not work but explaining the report is beyond the scope of this text.

Other options can be passed to the symphony solver such as `time_limit`, `gap_limit`, and `first_feasible` which all allow for optimization runs to be finished before it has solved the model to known optimality. These early termination options can be very helpful when working with large integer programming problems.

The `time_limit` option has a default value of -1 meaning no time limit. It should be an integer to indicate the number of seconds to run before stopping.

The `node_limit` option has a default value of -1 meaning no limit on the number of nodes (in an MIP problem, the number of linear programs). It should be an integer to indicate the number of nodes examined before stopping.

The last option, first_feasible, has a default value of FALSE. If it is set to TRUE, then symphony will stop when it has found a first solution that satisfies all the constraints (including integrality) rather than continuing on to prove optimality.

Let's examine the impact of just looking at the first feasible solution found by passing first_feasible=TRUE to see what happens.

```
tic("symphony")
result_symphony_ff <- solve_model(
  model, with_ROI(solver = "symphony",
                  verbosity = -1, first_feasible=TRUE))
symphonytimeff <- toc()
```

```
## symphony: 33.53 sec elapsed
```

```
print(solver_status(result_symphony_ff))
```

```
## [1] "infeasible"
```

```
symphonytimeff <- c("symphony first feas.",
                    symphonytimeff$toc - symphonytimeff$tic,
                   result_symphony_ff$status,
                   result_symphony_ff$objective_value)
```

Several interesting and important things should be noted here. First, ompr status indicates that the problem is infeasible. Just as we saw in Chapter 2's unbounded case, ompr interprets the status code from the solver by way of ROI as not being solved to optimality and concludes that the problem is not feasible.

This is a known issue in ompr and ompr.ROIas of version 0.8.1. It highlights that "infeasible" status from the ompr should be thought of as meaning that an optimal solution was not found for *some* reason such as being told to terminate after the first feasible solution was found, time limit reached, node limit reached, the MIP was infeasible, the MIP was unbounded, or some other issue.

A second thing to note that in my randomly generated instance, the very first feasible solution it found, happens to have the same objective function value as the optimal solution found to optimality earlier by glpk and symphony.

This is similar to the very simple, two variable integer programming problem that we examined using a branch and bound tree in the previous chapter but stopped before searching down the rest of the tree. Symphony just doesn't have confirmation yet that this first feasible solution is truly optimal yet.

Thirdly, on my computer it sometimes took less time to solve to optimality than for when it stopped *early* with just the initial feasible solution.

This demonstrates the variability of solving time and that while the warehouse optimization problem has quite a few variables, in its current structure, it is not a particularly difficult integer programming problem.

7.3.3 Performance of `lpsolve`

The `lpsolve` package has a long history too and is widely used. Let's test it and see how it performs.

```
tic("lpsolve")
result_lpsolve <-  solve_model(
  model, with_ROI(solver = "lpsolve", verbose=FALSE))
lpsolvetime <- toc()
```

```
## lpsolve: 27.28 sec elapsed
```

```
print(solver_status(result_lpsolve))
```

```
## [1] "optimal"
```

```
lpsolvetime1 <- c("lpsolve",
                  lpsolvetime$toc - lpsolvetime$tic,
                   result_lpsolve$status,
                   result_lpsolve$objective_value)
```

Again, we see that `lpsolve` was also successful.

7.3.4 Performance of gurobi

Gurobi is a comprehensive commercial optimization solver and is generally considered to be the fastest mixed integer programming solver. It is widely used for large scale commercial applications such as Amazon, Starbucks, and others. It can be run through a variety of platforms including C, Python, AMPL, GAMS, and other platforms. It is also available through R but requires more work than the open solvers that have previously used. To use Gurobi in R, do the following:

1. Get a license to gurobi (free licenses are available for academic use)
2. Download and install the gurobi software
3. Follow directions to have software access license key
4. Install gurobi's R package. Note that the package file is not in CRAN and will need to be installed directly from the gurobi installed files on your local computer. (For example in my windows install, it was in `C:\\gurobi900\win64\R` directory.)
5. Install the `roi.plugin.gurobi` package. Note that this may need to be installed from a github account as it is not officially supported by Gurobi.

A code chunk is provided without evaluation (by setting a code chunk option of `eval=FALSE`) for several reasons:

- Avoid issues of reproducibility for people that don't have access to gurobi
- Avoid potential of breaking working code due license expiration
- Maintain the philosophy of open-source tools
- Performance testing of gurobi under these circumstances might not do justice to the performance benefits of gurobi on much more complex optimization problems.

```
tic("gurobi")
result_gurobi <-  solve_model(
  model, with_ROI(solver = "gurobi"))
gurobitime <- toc()
print(solver_status(result_gurobi))
gurobitime1 <- c("gurobi",
                  gurobitime$toc - gurobitime$tic,
                   result_gurobi$status,
                   result_lpsolve$objective_value)
```

The `ROI.plugin` for each respective solver general accepts a subset of all the parameters or options that a solver can use. This may mean some trial and error is involved. The Gurobi documentation[7] provides a comprehensive list of parameters.

I've tested the following commonly used parameters with Gurobi and can confirm that they appear to work.

- `TimeLimit` terminates solving based on seconds, (usage: `TimeLimit=40.0`)
- `SolutionLimit` limits the number of MIP feasible solutions, (usage: `SolutionLimit=20`)
- `NodeLimit` limits the number of MIP nodes, only used in MIP, (usage: `NodeLimit=200`)
- `BestObjStop` stops when a solution is as good as value, (usage: `BestObjStop=50000.0`)
- `BestBdStop` stops when best bound is as good as value (usage: `BestBdStop=50000.0`)
- `MIPGap` stops based on relative gap between best bound and current solution. Note that the default is 10^{-4} rather than 0. A value of 0.1 effectively means that a solution is accepted and solving terminated if it is known that no solution is possible that is 10% better. (usage: `MIPGap=0.1`)
- `MIPGapAb` stops based on absolute gap between best bound and current solution. For example if the objective function is in dollars, a value of 1000.0 means that a solution is accepted and solving terminated if it is known that no solution is possible is more $1000 better. (usage: `MIPGap=1000.0`)

The gurobi solver has far more capabilities than are presented here. It provides it's own algebraic modeling language and as an integrated solution is able to avoid the issue of misinterpreting solving codes. For people with very large scale optimization problems that they are solving on a frequent basis, it may be worth investigating these other options. The `ompr` tool and algebraic modeling language used within R will make the transition straightforward.

7.3.5 Comparing Results across Solvers

Now let's compare how three major open source optimization solvers performed.

```
timeresults <- rbind(glpktimel, symphonytimel,
                     symphonytimeff, lpsolvetimel)
colnames (timeresults) <- c("Solver", "Time (Sec)",
                            "Status",
```

[7] https://www.gurobi.com/documentation/9.0/refman/parameters.html#sec:Parameters

```
                                    "Obj. Func. Value")
rownames (timeresults) <- NULL

kbl (timeresults, booktabs = T, digits=2,
    caption="Comparison of Results from Warehouse
       Customer Assignment Across Solvers") |>
  kable_styling(latex_options = "hold_position")
```

TABLE 7.7 Comparison of Results from Warehouse Customer Assignment Across Solvers

Solver	Time (Sec)	Status	Obj. Func. Value
glpk	31.8	optimal	60736
symphony	32.33	optimal	60736
symphony first feas.	33.53	infeasible	60736
lpsolve	35.43	optimal	60736

The most important thing to note is that each solver states that it found the same value for an objective function. While they may vary in how it is achieved if there are multiple optima (in other words, different values for decision variables x and y), this means that they all found a correct solution. Of course it would be a big problem if they claimed to have optimal solutions with different objective function values.

Note that time will vary significantly from computer to computer and may also depend upon the other tasks and the amount of memory available. Performance will change for different solver settings and for different problems. Also, performance time will change for different random instances – simply changing the random number seed will create different run-time results. This problem is probably too small to show off the advantages of gurobi and running gurobi through third party tool interfaces such as `ompr` and `ROI` will also be slower than a more native format.

The ROI package lists 19 different solvers with ROI plugins. Three commercial solvers, C-Plex, GUROBI, and MOSEK are worth highlighting. These programs may be substantially faster on large MIP problems and may make versions available for academic use. Others are for quadratic optimization or special classes of optimization problems.

After running these chunks dozens of times and finding the results to be quite stable across the various solvers tested, using a code chunk option `cache=TRUE` may be beneficial. Caching means that it will save and reuse the results unless things are changed. If it determines that things have changed, it will rerun that chunk. This can save a lot of time in knitting the chapter or rebuilding the book. This is a good application area for demonstrating the benefits of using the chunk option for caching. The current application requires about a minute to do each full size customer warehouse optimization run. Caching just the four runs can save a lot of knitting time. RStudio and knitr do a sophisticated check for changes which can invalidate the previously cached results and will then rerun the code chunk and save the results for future cached knits as appropriate. More demanding needs can use the granularity of numerical options for caching as compared to simply TRUE/FALSE and additional options. These might be required in detailed performance testing or numerical performance comparisons.

While caching can be very helpful with large runs, I have also run into difficulty where data or a code chunk has changed in a way that I know should affect results and yet the results didn't change. In fact, when running all the code chunks, the correct results would be shown. After too long a time spent debugging, I realized that it was a problem with the knitted file using old cached results rather than actually rerunning the current results. Technically, I had expected the previous cached results to be considered invalidated. In general, I would recommend not using knitr's caching of results unless computation time or resources become a problem and the user is willing to learn a little about the issue of caching.

Significant computer time can be used for large models but a variety of questions can be examined such as the following.

- How is problem size related to solving time?
- Are there systematic differences in speed or solving success for different solvers? (ex. does Gurobi run faster?)
- How do various modeling parameters affect solution speed?

7.3.6 Popularity of LP Solvers

Let's take another look at comparing solvers by examining their popularity. More precisely, let's examine the download logs of CRAN for a variety of linear programming solvers.

Many of these packages are being used directly in other packages. While we can't see which packages are being specifically used most often for `ompr`, we can see which ROI plugins are downloaded most often, which is probably a good proxy.

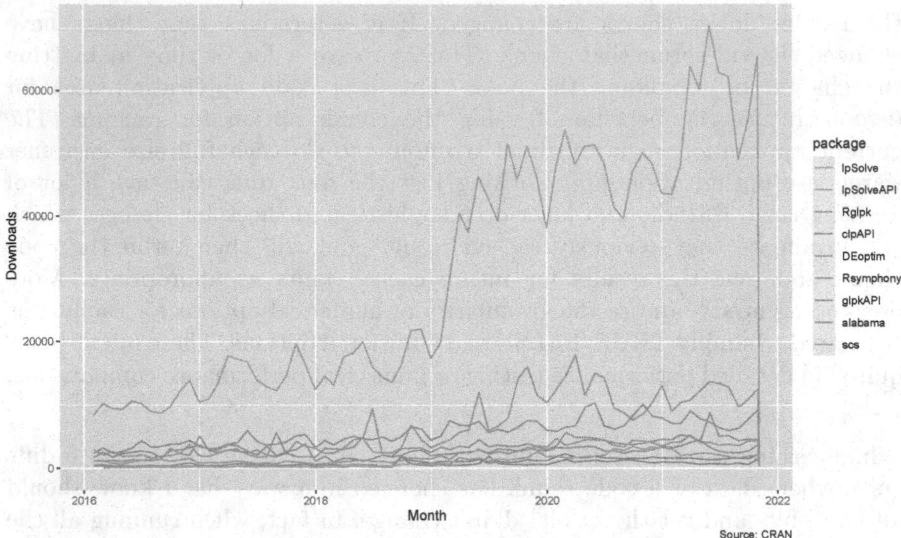

FIGURE 7.4 Monthly Downloads of Optimzation Packages.

We will focus our attention on the ROI packages that emphasize linear programming. This means dropping packages such as `ROI.plugin.nloptr` and `ROI.plugin.quadprog`.

Also, commercial software that is not available on CRAN such as gurobi and XPress, are not included.

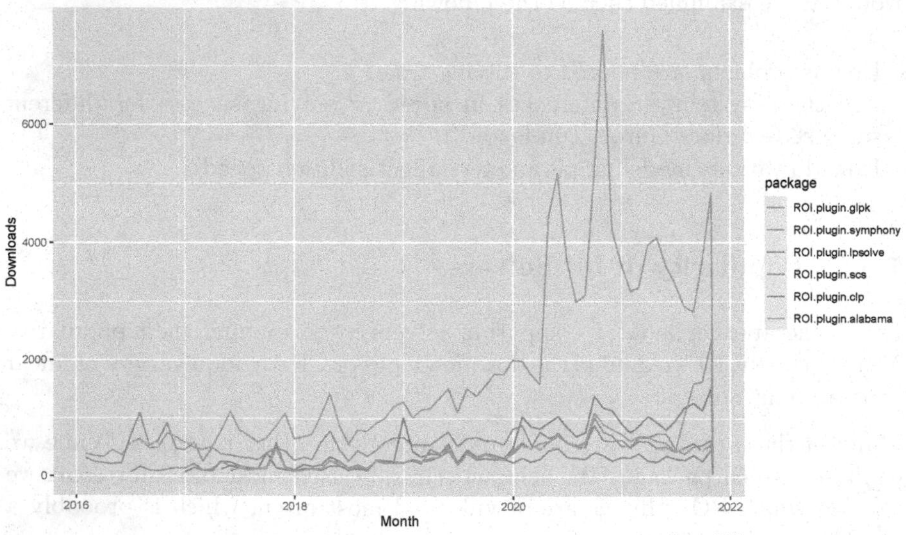

FIGURE 7.5 Monthly Downloads of Major ROI LP Plugins.

It is interesting to see that the symphony and glpk ROI plugins were relatively equal in downloads until about 2015 when glpk started to open a lead in usage. This of course does not mean that glpk is better than the others but is interesting to note. The spikes in downloads could be driven by other packages that use this particular solver, visibility from vignettes, courses, other publications, or package updates.

```
pkg_dl_data <- cran_stats(packages = c("ompr"))
ggplot(pkg_dl_data, aes(end, downloads, group=package,
                        color=package)) +
    geom_line() +
    labs(subtitle = "ompr:  Optimization Modeling Package for R",
      caption = "Source: CRAN",
      x = "Month", y = "Downloads")
```

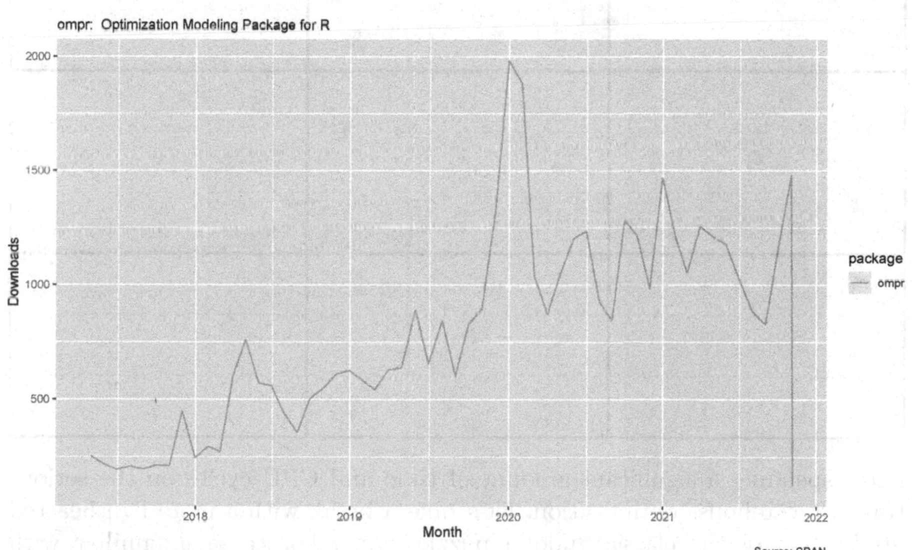

FIGURE 7.6 Monthly Downloads of ompr.

Let's close this section by examining the popularity of ompr itself.

Clearly a wide range of optimization tools can be used. Let's return to optimization formulations with a more entertaining application.

7.4 Solving Sudoku Puzzles using Optimization

7.4.1 Introduction to Sudoku and Optimization

Would you like to play a game? – Joshua (Wargames-1983)

				9				
							4	
			6		5			
						9		
	9	8	5					
			1	6			7	8
1		5	9		2			
			3					
9					4		3	

After spending a significant amount of time and CPU cycles on the serious topic of warehouse optimization, let's take a break with a more lighthearted application of the classic Sudoku puzzle game. For those unfamiliar with Sukoku, the core idea is to fill a 9 by 9 grid, using every digit from 1 to 9 once in each column, once in each row, and once in each 3x3 cluster. A limited number of "hints" are provided in terms of digits that are shown, often around 20 to 30 digits. Feel free to work on this Sudoku puzzle hand.

Using optimization, you can solve this or any other Sudoku model!

This section is based on an article by Dirk Schumacher and helper functions from the 'SudokuAlt' package from Bill Venables, one of the Godfathers of R. Let's start with Dirk's explanation, with permission, of the model.

> In this vignette we will solve Sudoku puzzles using MILP. Sudoku[8] in its most popular form is a constraint satisfaction problem and by setting the objective function to 0 you transform the optimization problem into a pure constraint satistication problem. In this document we will consider Sudokus in a 9x9 grid with 3x3 sub-matrices.
> Of course you can formulate an objective function as well that directs the solver towards solutions maximizing a certain linear function.
> The idea is to introduce a binary variable x with three indexes i, j, k that is 1 if and only if the number k is in cell i, j.

7.4.2 Formulating the Sudoku Problem

The basic optimization formulation is interesting for a few reasons.

First, there isn't a clear objective function. We are simply trying to find a a set of values that solves the problem without violating any of the rules of Sudoku. In other words, any feasible solution would be acceptable. As Dirk Schumacher stated, this is technically a constraint satisfaction problem rather than optimization because there is no objective function to optimize. Having said that, our tools of optimization are still useful.

We could minimize $x_{1,1,1}$, the sum of all decision variables (which would be 81, corresponding to the number of cells in the Sudoku grid.) We could also just tell it to maximize a particular number, say 42, or minimize another number like 0, which is exactly what Dirk did.

In building our constraints, it is easy to get overwhelmed as we have not used a triple subscripted variable before. Let's start by focusing on just the top left corner cell ($i = 1; j = 1$). If this cell were to have the digit 1, then that would mean $x_{1,1,1} = 1$. If that cell holds a 2, then $x_{1,1,2} = 1$. If it holds a 3, then $x_{1,1,3} = 1$ and so on up to the ninth digit, $x_{1,1,9} = 1$. We know that it cell must have one and only one digit. Therefore, for this top left cell, we

[8] https://en.wikipedia.org/wiki/Sudoku

can just add up all of the digit possibilities for that cell and set it equal to one.

$$\begin{aligned} & x_{1,1,1} + x_{1,1,2} + x_{1,1,3} + \\ & x_{1,1,4} + x_{1,1,5} + x_{1,1,6} + \\ & x_{1,1,7} + x_{1,1,8} + x_{1,1,9} \qquad = 1 \end{aligned}$$

Rather than writing out each possibility, we can instead use a summation for every digit, k.

$$\sum_{k=1}^{9} x_{1,1,k} = 1$$

This would still entail writing a similar constraint for each of the other 80 cells such as $(i = 1; j = 2)$, $(i = 1; j = 3)$, and so on. Instead, this is a perfect opportunity to use the $\forall$ to repeat this for all values of i and j.

$$\sum_{k=1}^{9} x_{i,j,k} = 1, \ \forall i, j$$

We then need to ensure that in each row, i, each digit k only appears in one of the columns.

$$\sum_{i=1}^{9} x_{i,j,k} = 1, \ \forall j, k$$

We then need to ensure that in each column, j, each digit only appears in of the rows.

$$\sum_{j=1}^{9} x_{i,j,k} = 1, \ \forall i, k$$

Next, we need to create a constraint for each 3x3 grouping so that it contains each digit only once. This is more complicated. Let's consider the top left 3x3 cluster. For each digit k, we need to ensure that only one cell has it.

$$\begin{aligned} & x_{1,1,k} + x_{1,2,k} + x_{1,3,k} + \\ & x_{2,1,k} + x_{2,2,k} + x_{2,3,k} + \\ & x_{3,1,k} + x_{3,2,k} + x_{3,3,k} \qquad = 1, \ \forall \ k \end{aligned}$$

Let's generalize this by using summations.

$$\sum_{i=1}^{3}\sum_{j=1}^{3} x_{i,j,k} = 1, \; k = 1, ..., 9$$

Unfortunately this set of constraints only covers one cluster.

If we are clever, we may note that the top left corner of each cluster is a row or column value of 1, 4, or 7.

We could extend this constraint to handle each set of clusters. We'll use a counter across for each cluster by row, using r. Similarly, we'll use c for cluster.

$$\sum_{i=1+3r}^{3+3r}\sum_{j=1+3c}^{3+3c} x_{i,j,k} = 1, \; k = 1, ..., 9 \; r = 0, 1, 2, \; c = 0, 1, 2,$$

We can now pull everything together into a single formulation. Again, remember that we could use *any* objective function since we are simply trying to find a feasible solution. We will just picking maximizing the number zero since ompr defaults to maximization as an objective function. We could also tell a solver such as symphony to just find a first feasible solution and terminate.

$$
\begin{aligned}
\max \;\; & 0 \\
\text{s.t.:} \;\; & \sum_{k=1}^{9} x_{i,j,k} = 1 && i = 1, ..., n, \; j = 1, ..., n \\
& \sum_{i=1}^{n} x_{i,j,k} = 1 && j = 1, ..., n, \; k = 1, ..., 9 \\
& \sum_{j=1}^{n} x_{i,j,k} = 1 && i = 1, ..., n, \; k = 1, ..., 9 \\
& \sum_{i=1+r}^{3+r}\sum_{j=1+c}^{3+c} x_{i,j,k} = 1, && k = 1, ..., 9 \; r = 0, 3, 6, \; c = 0, 3, 6, \\
& x_{i,j,k} \in \{0, 1\} && i = 1, ..., n, \; j = 1, ..., n, \; k = 1, ..., 9
\end{aligned}
$$

7.4.3 Implementing Sudoku in ompr

We are now ready to implement the model. As always, clearly defining variables and constraints is important. A triple subscripted variable often makes for a tricky model to both formulate and implement. Also, that last constraint may take careful examination.

```
n <- 9
Sudoku_model <- MIPModel() |>

  # The number k stored in position i,j
  add_variable(Vx[i, j, k], i = 1:n, j = 1:n, k = 1:9,
               type = "binary")    |>

  # no objective
  set_objective(0)                 |>

  # only one number can be assigned per cell
  add_constraint(sum_expr(Vx[i, j, k], k = 1:9) == 1,
                 i = 1:n, j = 1:n) |>

  # each number is exactly once in a row
  add_constraint(sum_expr(Vx[i, j, k], j = 1:n) == 1,
                 i = 1:n, k = 1:9) |>

  # each number is exactly once in a column
  add_constraint(sum_expr(Vx[i, j, k], i = 1:n) == 1,
                 j = 1:n, k = 1:9) |>

  # each 3x3 square must have all numbers
  add_constraint(sum_expr(Vx[i, j, k], i = 1:3 + r,
                          j = 1:3 + c) == 1,
                 r = seq(0, n - 3, 3),
                 c = seq(0, n - 3, 3), k = 1:9)
Sudoku_model
```

```
## Mixed integer linear optimization problem
## Variables:
##   Continuous: 0
##   Integer: 0
##   Binary: 729
## Model sense: maximize
## Constraints: 324
```

We will use `glpk` to solve the above model. Note that we haven't fixed any numbers to specific values. That means that the solver will find a valid Sudoku without any prior hints.

I've made a couple of minor changes to Dirk's code chunks. In the implementation, I replaced the x variable in ompr with Vx to avoid R name space collisions with previously defined R data structures named x in my environment. Secondly, I switch the sx and sy to r and c to represent moving over by rows and columns in the tricky last constraint for clusters.

```
Sudoku_result <- solve_model(
  Sudoku_model, with_ROI(
    solver = "glpk", verbose = TRUE))
```

```
## <SOLVER MSG>  ----
## GLPK Simplex Optimizer, v4.47
## 324 rows, 729 columns, 2916 non-zeros
##       0: obj =  0.000000000e+000  infeas = 3.240e+002 (324)
##     500: obj =  0.000000000e+000  infeas = 2.000e+001 (79)
## *   520: obj =  0.000000000e+000  infeas = 4.644e-014 (75)
## OPTIMAL SOLUTION FOUND
## GLPK Integer Optimizer, v4.47
## 324 rows, 729 columns, 2916 non-zeros
## 729 integer variables, all of which are binary
## Integer optimization begins...
## +   520: mip =     not found yet <=              +inf        (1; 0)
## +  1414: >>>>>  0.000000000e+000 <=  0.000000000e+000   0.0% (43; 0)
## +  1414: mip =  0.000000000e+000 <=     tree is empty   0.0% (0; 85)
## INTEGER OPTIMAL SOLUTION FOUND
## <!SOLVER MSG> ----
```

```
# the following dplyr statement plots a 9x9 matrix
Solution <- Sudoku_result |>
  get_solution(Vx[i,j,k]) |>
  filter(value > 0) |>
  select(i, j, k) |>
  tidyr::spread(j, k) |>
  select(-i)
```

If you want to solve a specific Sudoku you can fix certain cells to specific values. For example here we solve a Sudoku that has the sequence from 1 to 9 in the first 3x3 matrix fixed.

```
Sudoku_model_specific <- Sudoku_model |>
  add_constraint(Vx[1, 1, 1] == 1) |>
     # Set digit in top left to the number 1
  add_constraint(Vx[1, 2, 2] == 1) |>
     # Set digit in row 1, column 2 to the number 2
  add_constraint(Vx[1, 3, 3] == 1) |>
     # Set digit in row 1, column 3 to the number 3
  add_constraint(Vx[2, 1, 4] == 1) |>
     # Set digit in row 2, column 1 to the number 4
  add_constraint(Vx[2, 2, 5] == 1) |>  # etc....
  add_constraint(Vx[2, 3, 6] == 1) |>
  add_constraint(Vx[3, 1, 7] == 1) |>
  add_constraint(Vx[3, 2, 8] == 1) |>
  add_constraint(Vx[3, 3, 9] == 1)
Sudoku_result_specific <- solve_model(
  Sudoku_model_specific, with_ROI(
    solver = "glpk", verbose = TRUE))
```

```
## <SOLVER MSG>  ----
## GLPK Simplex Optimizer, v4.47
## 333 rows, 729 columns, 2925 non-zeros
##       0: obj =  0.000000000e+000  infeas = 3.330e+002 (333)
## *   477: obj =  0.000000000e+000  infeas = 2.015e-015 (84)
## OPTIMAL SOLUTION FOUND
## GLPK Integer Optimizer, v4.47
## 333 rows, 729 columns, 2925 non-zeros
## 729 integer variables, all of which are binary
## Integer optimization begins...
## +   477: mip =     not found yet <=              +inf        (1; 0)
## +   980: >>>>>  0.000000000e+000 <=  0.000000000e+000   0.0% (31; 0)
## +   980: mip =  0.000000000e+000 <=     tree is empty   0.0% (0; 61)
## INTEGER OPTIMAL SOLUTION FOUND
## <!SOLVER MSG> ----
```

```
Solution_specific <- Sudoku_result_specific |>
  get_solution(Vx[i,j,k]) |>
  filter(value > 0) |>
  select(i, j, k) |>
  tidyr::spread(j, k) |>
  select(-i)
```

```
kbl (Solution_specific)   # Display solution to Sudoku Puzzle
```

1	2	3	4	5	6	7	8	9
1	2	3	8	4	6	9	5	7
4	5	6	1	9	7	3	8	2
7	8	9	3	5	2	1	4	6
2	3	4	7	1	8	6	9	5
8	9	1	5	6	3	2	7	4
5	6	7	4	2	9	8	3	1
3	4	5	6	8	1	7	2	9
6	7	2	9	3	4	5	1	8
9	1	8	2	7	5	4	6	3

Now, let's try printing it nicely using the `sudokuAlt` package from Bill Venables, available from CRAN.

```
plot(as.sudoku(as.matrix(Solution_specific, colGame="grey")))
```

1	2	3	8	4	6	9	5	7
4	5	6	1	9	7	3	8	2
7	8	9	3	5	2	1	4	6
2	3	4	7	1	8	6	9	5
8	9	1	5	6	3	2	7	4
5	6	7	4	2	9	8	3	1
3	4	5	6	8	1	7	2	9
6	7	2	9	3	4	5	1	8
9	1	8	2	7	5	4	6	3

Have at it. Any Sudoku puzzle can be solved using this model as long as there is a feasible solution. Note that if there are two or more solutions to the puzzle based on the hints, this model will find one of them and does not indicate whether there might be other solutions.

Feel free to time the solution using the `tictoc` package.

The intention here is not to imply that optimization is the only way or the best way to solve Sudoku problems. There are algorithmic approaches for solving Sudoku that can be more computationally efficient than using a general purpose integer programming system and in fact people have implemented such algorithms in R and other languages. The purpose of this example was to show how a nontraditional problem can be framed and implemented.

7.4.4 Sudoku Discussion

In this chapter, we covered some problems with medium-sized integer programming problems. The same approaches and models can be scaled up to larger problems.

Improvements in software and computer hardware have opened up optimization to a variety of large scale problems. Integrating this into R has made it straightforward to connect together a variety of tools including simulation and data visualization.

7.5 Exercises

Even moderate sized integer programming problems can result in significant computational loads. While these problems will not represent a heavy load for a desktop or laptop computer, they be intensive for netbooks or cloud-based services. If you are using a cloud-based service such as RStudio.cloud, a server, or third party service such as Amazon Web Services, be sure to understand the implications of heavy computational loads. Possible results of excessive use of third party computer time include throttling of access similar to cell phone users exceeding monthly data caps, high "overage" charges, a warning from a systems administrator, or even account termination.

Exercise 7.1 (Warehouse-Max Customers-Small). Run the analysis for the smaller warehouse optimization but with just 10 customers and 3 warehouses. Extend the warehouse customer assignment problem to have each warehouse that is built be assigned a maximum of 4 customers. Compare the solution to that obtained earlier. Ensure that your data for customers and warehouses is the same for the two cases.

This problem size should be cloud-friendly but ensure that you are solving the appropriate sized problem before running.

Exercise 7.2 (Warehouse-Max and Min Customers-Small)**.** Run the analysis for the smaller warehouse optimization but with just 10 customers and 3 warehouses. Extend and solve the warehouse customer assignment problem to have each warehouse that is built be assigned a maximum of 4 customers and a minimum of 2 customers. Show plots of both solutions. Compare the solution to that obtained earlier using tables as appropriate. Ensure that your data for customers and warehouses is the same for the two cases.

This problem size should be cloud-friendly but ensure that you are solving the appropriate sized problem before running.

Exercise 7.3 (Warehouse-Simulating Customers-Moderate)**.** Run the analysis for the smaller warehouse optimization with a max of 10 customers and 3 warehouses. Simulate it 20 times and plot the results in terms of number of warehouses used and the total cost. Discuss the results.

This problem size should not be be run on a cloud or remote server without full understanding of load implications. Note that this will be somewhat computationally intensive and is best done on a local computer rather than the cloud. Doing it on a personal computer may require on the order of five minutes of processing time.

Exercise 7.4 (Warehouse-Max Customers-Big)**.** Extend the warehouse customer assignment problem to have each warehouse that is built be assigned a maximum of 40 customers. Compare the solution to that obtained earlier. Ensure that your data for customers and warehouses is the same for the two cases. This problem size should not be be run on a cloud or remote server without full understanding of load implications.

Exercise 7.5 (Warehouse-Max and Min Customers-Big)**.** Extend the warehouse customer assignment problem to have each warehouse that is built be assigned a maximum of 40 customers and a minimum of 15 customers. Compare the solution to that obtained earlier. Ensure that your data for customers and warehouses is the same for the two cases. This problem size should not be be run on a cloud or remote server without full understanding of load implications.

Exercise 7.6 (Warehouse-Simulating Customers-Big)**.** Run the analysis for the warehouse optimization model 20 times and plot the results in terms of number of warehouses used and the total cost. Discuss and interpret the results. Note that this will be **computationally intensive** and should not be done on the cloud. Doing it on a personal computer may require on the order of five minutes of processing time.

Exercise 7.7 (Sudoku-First Puzzle)**.** Solve the Sudoku puzzle from the beginning of this section using R.

Exercise 7.8 (Sudoku-Bottom Row)**.** Solve a Sudoku model using optimization where the bottom row is 9 numbered down to 1.

Exercise 7.9 (Sudoku-Right Column)**.** Solve a Sudoku model using optimization where the rightmost column is 9 numbered down to 1.

Exercise 7.10 (Sudoku-Center Cluster)**.** Solve a Sudoku model using optimization where the center 3x3 cluster is made up of the numbers from 9 to 1. (First row is 9, 8, 7; second row is 6, 5, 4; bottom row is 3, 2, 1)

Exercise 7.11 (Find a Sudoku to Solve)**.** Find a Sudoku puzzle from a newspaper or elsewhere and solve it using optimization.

Exercise 7.12 (Sudoku-First Puzzle)**.** Solve the Sudoku problem from the beginning of this section.

				9				
							4	
			6		5			
						9		
	9	8	5					
			1	6			7	8
1		5	9		2			
			3					
9					4		3	

Exercise 7.13 (Sudoku-Multiple Optima)**. Challenge:** Sudoku puzzles are created with the intention of having a single solution since the approaches

people use to solve them are based on incrementally reasoning out what values each cell must contain. The result is that the ambiguity of multiple optima may cause problems for people to solve. While the optimization model can solve these without difficulty, it is only going to find one, arbitrary solution. Simply resolving is likely to give you the same solution even when there are multiple optima. Consider variations and apply them to find multiple solutions to a problem. For an example of one with multiple optimal solutions, the most extreme case would be starting with a blank board.

Exercise 7.14 (Sudoku-Infeasible). **Challenge:** A Sudoku puzzle with only a few numbers shown can be infeasible even though it does not contain any obvious violation of repeated digits. People building Sudoku puzzles need to ensure that every puzzle is solvable or risk having frustrated customers. Add a digit one at a time to an existing Sudoku puzzle board that does not immediately violate the repeated digits requirement, test for a valid solution, and repeat, until the Sudoku game board becomes infeasible.

Exercise 7.15 (Sudoku-4x4). **Ultra-Challenge:** Solve a 16 letter Sudoku puzzle using R. It consists of 4x4 grids and uses the letters starting with A in place of numbers.

G									E				L		
			K				I							C	O
		F								C	O		B	J	
		F	M	K			J	L	B	D				A	P
		D		L											
		K	N						A	L			G		E
						J	N		K						
	I							N		O		L			
K												J			
								J					K		
		J					K						D		
P		N	I	J			L								
				F		K		C		N					
C							M								
	D			A	C					K		N			M
		P		E		D			G						

7.6 Production Planning over Time

Let's revisit our drone manufacturing example but accounting for time. In this case, we'll look at producing ants and bats over time while accounting for with varying weekly demand, inventory carrying costs, setup costs, marginal production costs, and beginning inventory. The planning horizon is four weeks.

Let's start by by showing the data.

```
nTime <- 4
mDemandA <- 150  # Used for generating demand
mDemandB <- 250  #  Upper limit for random number generator

demA <- c(round(runif(nTime)*mDemandA))
demB <- c(round(runif(nTime)*mDemandB))

beg_invA <- 71; beg_invB <- 98  # Inv at beginning of week 1

fcostA <- 80; fcostB <- 140  # Setup cost for product

invcostA <- 1.5; invcostB <- 2.0 # Weekly carry cost per unit

prodcostA <- 10; prodcostB <- 11 # Marginal production cost of A

lmaxinv <- 400  # Maximum combined inventory
lmaxpro <- 600  # Maximum combined production
```

Our demand over time for Ants and Bat is then the following.

TABLE 7.8 Product Demand Over Time

	Week 1	Week 2	Week 3	Week 4
demA	120	45	133	128
demB	209	18	156	128

Our goal is to create a production plan that will meet all demand at the lowest total cost.

Let's start by defining our data. Instead of Ants and Bats, we'll abbreviate it to just A and B.

Data:

- C^I_A = The weekly inventory carrying cost for each Ant carried from one week to the next.

- C_B^I = The weekly inventory carrying cost for each Bats carried from one week to the next.
- C_A^P = The marginal production cost for each Ant.
- C_B^P = The marginal production cost for each Bat.
- C_A^S = The setup cost for producing Ants in any given week.
- C_B^S = The setup cost for producing Bats in any given week.
- B_A^I = Inventory of Ants at the beginning of week 1.
- B_B^I = Inventory of Bats at the beginning of week 1.
- $D_{t,A}$ = Demand for Ants in week t.
- $D_{t,B}$ = Demand for Bats in week t.
- L^P = Maximum Limit on production of Ants and Bats in any given week.
- L^I = Limit on maximum inventory of Ants and Bats in any given week.

Decision variables:

- $x_{t,A}$ is the number of Ants to make it week t.
- $x_{t,B}$ is the number of Bats to make it week t.
- $y_{t,A} = 1$ if a setup for Ant production is done in week t; 0 otherwise.
- $y_{t,B} = 1$ if a setup for Bat production is done in week t; 0 otherwise.
- $z_{t,A}$ is the inventory of Ants at the end of week t.
- $z_{t,B}$ is the inventory of Bats at the end of week t.

Let's start with some easy relationships. We know that we can't exceed total production capacity in any given week. All we need to do is add the two production variables in week t together and constrain it to not exceed the production capacity. For week 1, we would have the following:

$$x_{1,A} + x_{1,B} \leq L^P$$

We can then generalize this for all weeks in the following manner.

$$x_{t,A} + x_{t,B} \leq L^P, \ \forall t$$

We can then implement an upper limit on total inventory in the same way.

$$z_{t,A} + z_{t,B} \leq L^I, \ \forall t$$

A critical concept to understand is the inventory balance. Let's demonstrated it with Ants and arbitrarily pick week 3. We know that the ending inventory for week 3, $z_{3,A}$ is equal to the inventory from the from the previous week, $z_{2,A}$, plus amount produced in week 3, $x_{3,A}$, minus the amount sold (in our case demand), $D_{3,A}$.

$$z_{3,A} = z_{2,A} + x_{3,A} - D_{3,A}$$

The inventory balance constraints follow a similar form for each week other than week 1, so it is straightforward to generalize this to account for other time periods after the initial one, $t = 1$.

$$z_{t,A} = z_{t-1,A} + x_{t,A} - D_{t,A},\ \forall t > 1$$

For week 1, the previous inventory is fixed rather than a variable and can be framed as the following.

$$z_{1,A} = B_A + x_{1,A} - D_{1,A}$$

Next, we need to remember to create a linking constraint for each of decisions of how much to produce with the decision to produce in a given time period. In this case, we know that in any given time period, we have an upper limit on production, L^P so we will use that as our Big M value. For Ants, the constraint then follows this form for each week t.

$$x_{t,A} \leq L^P \cdot y_{t,A}\ \forall\, t$$

The linking constraint for Bats follows the same form.

$$x_{t,B} \leq L^P \cdot y_{t,B}\ \forall\, t$$

Our goal is to minimize the total cost. Let's look at the total cost with respect Ants. It is a combination production cost, setup costs, and carrying costs.

$$\sum_{t=1}^{4} C_t^P \cdot x_{t,A} + C_t^S \cdot y_{t,A} + C_t^I \cdot z_{t,A}$$

This was just the costs for Ants (A) over time. We can create a similar function for Bats (B) and then since these are costs, we want minimize the sum of the two giving us our objective function.

$$\begin{split} \min \ \sum_{t=1}^{4}(C_A^P \cdot x_{t,A} + C_A^S \cdot y_{t,A} + C_A^I \cdot z_{t,A} + \\ C_B^P \cdot x_{t,B} + C_B^S \cdot y_{t,B} + C_B^I \cdot z_{t,B}) \end{split}$$

Now, let's create the full formulation.

$$
\begin{aligned}
\min\ & \sum_{t=1}^{4}(C_A^P \cdot x_{t,A} + C_A^S \cdot y_{t,A} + C_A^I \cdot z_{t,A} + \\
& C_B^P \cdot x_{t,B} + C_B^S \cdot y_{t,B} + C_B^I \cdot z_{t,B}) \\
\text{s.t.:}\ & z_{1,A} = B_A + x_{1,A} - D_{1,A} \\
& z_{1,B} = B_B + x_{1,B} - D_{1,B} \\
& z_{t,A} = z_{t-1,A} + x_{t,A} - D_{t,A},\ \forall t > 1 \\
& z_{t,B} = z_{t-1,B} + x_{t,B} - D_{t,B},\ \forall t > 1 \\
& x_{t,A} + x_{t,B} \leq L^P,\ \forall t \\
& z_{t,A} + z_{t,B} \leq L^I,\ \forall t \\
& x_{t,A} \leq L^P \cdot y_{t,A}\ \forall t \\
& x_{t,B} \leq L^P \cdot y_{t,B}\ \forall t \\
& x_{t,A}\ x_{t,A},\ z_{t,A}\ z_{t,B} \geq 0,\ \forall\ t \\
& y_{t,A}, y_{t,B}\ \in \{0,1\}\ \forall\ t
\end{aligned}
$$

7.6.1 Implementing the Model

We'll start by initializing the model and creating the decision variables.

```
promod <- MIPModel()

promod <- add_variable (promod, VxA[tt], tt=1:nTime,
                        lb=0, type="continuous")
promod <- add_variable (promod, VxB[tt], tt=1:nTime,
                        lb=0, type="continuous")
   # Production volume for products A and B in each time period.

promod <- add_variable (promod, VyA[tt], tt=1:nTime, type="binary")
promod <- add_variable (promod, VyB[tt], tt=1:nTime, type="binary")
   # Setups incurred for each product in each time period.

promod <- add_variable (promod, VzA[tt], tt=1:nTime,
                        lb=0, type="continuous")
promod <- add_variable (promod, VzB[tt], tt=1:nTime,
                        lb=0, type="continuous")
   # Inventory at end of period for each product.
```

Now let's define the objective function. We are not selling prices

```
promod <- set_objective(promod, sum_expr(prodcostA * VxA[tt] +
                                         prodcostB * VxB[tt] +
                                         fcostA * VyA [tt] +
                                         fcostB * VyB [tt] +
                                         invcostA * VzA [tt] +
                                         invcostB * VzB [tt],
                                         tt=1:nTime),
                                         "min")
```

Let's start with our constraints for linking production, inventory, and demand for the first week. The inventory at the end of week 1 is simply the beginning inventory at the start of week 1 (`beg_invA`) plus the amount produced minus the demand.

Note that this assumes all demand must be satisfied in the same week. This model could be extended to allow for backlogged demand or other variations.

```
promod <- add_constraint (promod,
                          beg_invA + VxA[1] - demA [1] == VzA[1] )
promod <- add_constraint (promod,
                          beg_invB + VxB[1] - demB [1] == VzB[1] )
```

This handled the first week. Now we need to do the same for all of the following weeks. We take the ending inventory from the previous week, add the new production, and then subtract the demand to give us the ending inventory for both products.

```
promod <- add_constraint (promod, VzA[tt-1] + VxA[tt] -
                          demA [tt] == VzA[tt], tt = 2:nTime )

promod <- add_constraint (promod, VzB[tt-1] + VxB[tt] -
                          demB [tt] == VzB[tt], tt = 2:nTime )
```

Now we need to use our create our linking constraints to connect our production and decision to produce products. We will use the Big M method and the maximum production production value as our Big M value.

```
promod <- add_constraint (promod, VxA [tt] <= lmaxpro * VyA[tt],
                          tt = 1:nTime)
```

```
promod <- add_constraint (promod, VxB [tt] <= lmaxpro * VyB[tt],
                          tt = 1:nTime)
```

Now let's put a limit our joint inventory of products.

```
promod <- add_constraint (promod, VzA[tt] + VzB[tt] <= lmaxinv,
                          tt = 1:nTime)
```

Now let's put a limit our joint production capacity of products.

```
promod <- add_constraint (promod, VzA[tt] + VzB[tt] <= lmaxpro,
                          tt = 1:nTime)
```

At this point, let's look at the size of the problem.

```
promod
```

```
## Mixed integer linear optimization problem
## Variables:
##   Continuous: 16
##   Integer: 0
##   Binary: 8
## Model sense: minimize
## Constraints: 24
```

We are ready to solve.

```
prores <- solve_model(promod, with_ROI(solver = "glpk"))
prores$status
```

```
## [1] "optimal"
```

```
prores$objective_value
```

```
## [1] 8856.5
```

We could do some light processing of results for display purposes.

```
Sol <- rbind (
  t(as.matrix(as.numeric(get_solution (prores, VxA[tt])[,3]))),
  t(as.matrix(as.numeric(get_solution (prores, VxB[tt])[,3]))),
  t(as.matrix(as.numeric(get_solution (prores, VyA[tt])[,3]))),
  t(as.matrix(as.numeric(get_solution (prores, VyB[tt])[,3]))),
  t(as.matrix(as.numeric(get_solution (prores, VzA[tt])[,3]))),
  t(as.matrix(as.numeric(get_solution (prores, VzB[tt])[,3]))))
var_list <- c("$x_{t_A}$", "$x_{t_B}$", "$y_{t_A}$", "$y_{t_B}$",
              "$z_{t_A}$", "$z_{t_B}$")
Sol <- cbind(var_list, Sol)
colnames(Sol) <- c("Variable", "Week 1", "Week 2", "Week 3", "Week 4")
rownames(Sol) <- c("Ant Production", "Bat Production",
                   "Ant Setup", "Bat Setup",
                   "Ant Inventory", "Bat Inventory")
```

TABLE 7.9 Production Planning Over Time

	Variable	Week 1	Week 2	Week 3	Week 4
Ant Production	x_{t_A}	94	0	133	128
Bat Production	x_{t_B}	129	0	156	128
Ant Setup	y_{t_A}	1	0	1	1
Bat Setup	y_{t_B}	1	0	1	1
Ant Inventory	z_{t_A}	45	0	0	0
Bat Inventory	z_{t_B}	18	0	0	0

This production planning over time example can be further enhanced in many directions to tailor it to specific applications. Some of the ways that it can be enhanced include the following:

- Set varying production limits over time.
- Set varying minimum inventory levels over time to help guard against variation and stockouts.
- Allow for not satisfying all demand
- Allow for backlogged inventory.
- Generalizing for the number of products.
- Simulating varying demand over time for multiple production runs.

- Varying demand uncertainty in the future (higher spread in the future).
- Increasing the number of products.

7.7 Additional Exercises

Exercise 7.16 (Production Planning). Antenna Engineering has received an order for a key component of 5G base stations from a major telecommunications company. They must provide 250 products and they have 6 machines that can be used to produce these products. The machines have differing characteristics described in the following table.

TABLE 7.10 Production Planning

Machine	Fixed Cost	Variable Cost	Capacity
1	100	2.1	50
2	95	2.3	60
3	87	2.5	75
4	85	2.4	40
5	80	2.0	60
6	70	2.6	80

The goal is to develop a production plan at the lowest possible cost.

a. Formulate an explicit mathematical optimization model for this problem using LaTeX in RMarkdown.
b. Implement and solve your optimization model using `ompr`.
c. Interpret the solution to the problem.
d. Create your own unique variation.
e. Create your own variation of the above problem by adding a logical relationship between machines that changes the solution.
f. Explain the modification in terms of the application, implement the change in the model, solve the revised model.
g. Compare and discuss your solution with respect to the original solution.
h. Create a generalized algebraic model in LaTeX, implement in `ompr`, and solve.
i. Compare results to the explicit model.

Exercise 7.17 (Warehouse Optimization). The warehouse optimization example in chapter 7 had a solution resulting in three warehouses. Assume that you had been responsible for creating the plan and presented the recommendation to the Vice President of Supply Chain Management who stated that they don't want to pay for the construction of three warehouses and that at most two warehouses should be built.

a. Without changing the model and only relying on the information from the book, give your quick response as to how this would affect the solution. (Assume that the VP is unfamiliar with optimization.)
b. Another employee familiar with optimization overhears your conversation with the VP. Explain your answer succinctly in optimization terms.
c. Write the constraint(s) to reflect the VP's suggestion. Write the LaTeX constraint(s) in terms of variables and/or data from the warehouse example in Chapter 7. Explain in your own words how the constraint(s) implement the suggested modification.
d. *Challenge:* Solve the revised problem.

Exercise 7.18 (Chemical Factories). Similar to the Warehouse problem in this chapter, here we have Chemical Factories for which we need to decide the locations. We need to set up 2 factories with 100 residents in the town. The town government is trying to keep the factories at maximum possible distance from the residents to avoid any decrease the impacts and risk of mishaps.

A. Create an appropriate formulation. B. Taking the randomly generated setup costs for factories similar to warehouses, run the analysis to decide the location for these 2 chemical factories which are at maximum possible distance from all residents. C. Discuss the results.

8

Goal Programming

8.1 Introduction

Up until this point, we assumed that there would be a single, clear objective function. Often we have more complex situations where there are multiple conflicting objectives. In our earlier production planning case, we might have additional objectives besides maximizing profit such as minimizing environmental waste or longer term strategic positioning. In the case of our capital budgeting problem, we can envision a range of additional considerations beyond simple expected net present value maximization. In the warehouse site selection problem from Chapter 7, we can envision other considerations such as political placement in certain states, being prepared for future varying growth levels in different regions, and other issues which could further influence a simple minimum cost solution.

8.2 Preemptive Goal Programming

Let's begin with a simple example. Recall the example of multiple optima from Chapter 2. We had adjusted the characteristics of *Cats* so that there were multiple optima with different mixes of *Ants* and *Cats*. The LP solver found an initial solution that only produced Ants, and we had to force the LP solver to find the *Cats* oriented production.

$$\begin{aligned}
\text{Max } & 7 \cdot Ants + 12 \cdot Bats + 14 \cdot Cats \\
\text{s.t.: } & 1 \cdot Ants + 4 \cdot Bats + 2 \cdot Cats \leq 800 \\
& 3 \cdot Ants + 6 \cdot Bats + 6 \cdot Cats \leq 900 \\
& 2 \cdot Ants + 2 \cdot Bats + 4 \cdot Cats \leq 480 \\
& 2 \cdot Ants + 10 \cdot Bats + 4 \cdot Cats \leq 1200 \\
& Ants,\ Bats,\ Cats \geq 0
\end{aligned}$$

DOI: 10.1201/9781003051251-8

Let's reframe this problem where the company's primary goal is to maximize profit but the Cat product is more prestigious than the Ant product and emphasizing it will benefit the company in long term market positioning. The company doesn't want to hurt profit but holding everything else equal, wants to then maximize Cat production.

The first step is to make the primary goal more direct. The objective function is now a fourth variable, *Profit* that is a direct function of the other three variables. The following formulation would then be considered the first phase.

$$\begin{aligned}
\text{Max } & Profit \\
\text{s.t.: } & Profit = 7 \cdot Ants + 12 \cdot Bats + 14 \cdot Cats \\
& 1 \cdot Ants + 4 \cdot Bats + 2 \cdot Cats \leq 800 \\
& 3 \cdot Ants + 6 \cdot Bats + 6 \cdot Cats \leq 900 \\
& 2 \cdot Ants + 2 \cdot Bats + 4 \cdot Cats \leq 480 \\
& 2 \cdot Ants + 10 \cdot Bats + 4 \cdot Cats \leq 1200 \\
& Ants,\ Bats,\ Cats \geq 0
\end{aligned}$$

The optimal objective function value from this LP is 1980 as shown in Chapter 2. Now we modify the above formulation to keep the *Profit* fixed at 1980 and maximize the production of *Cats*.

$$\begin{aligned}
\text{Max } & Cats \\
\text{s.t.: } & Profit = 7 \cdot Ants + 12 \cdot Bats + 14 \cdot Cats \\
& 1 \cdot Ants + 4 \cdot Bats + 2 \cdot Cats \leq 800 \\
& 3 \cdot Ants + 6 \cdot Bats + 6 \cdot Cats \leq 900 \\
& 2 \cdot Ants + 2 \cdot Bats + 4 \cdot Cats \leq 480 \\
& 2 \cdot Ants + 10 \cdot Bats + 4 \cdot Cats \leq 1200 \\
& Profit = 1980 \\
& Ants,\ Bats,\ Cats \geq 0
\end{aligned}$$

Rather than directly stating the solution value of 1980, a more general approach would use the term, $Profit^*$ to denote the optimal solution from the first Phase and the new Phase 2 constraint would then be $Profit = Profit^*$. This would then give a revised solution that emphasizes Cat production in so far as it doesn't detract from Profit. In other words, maximizing Profit preempts maximizing Cat production.

8.3 Policies for Houselessness

Let's use an example that is a pressing issue for many cities – homelessness. Note that is often better characterized as houselesness and the term will be used interchangeably for the sake of this illustration.

The City of Bartland has a problem with houselessness. Two ideas have been proposed for dealing with the houselessness problem. The first option is to build new, government subsidized tiny homes for annual cost of $10K$ which would serve one adult 90% of the time and a parent with a child 10% of the time. Another option is to create a rental subsidy program which costs $25K$ per year per unit which typically serves a single adult (15%), two adults (20%), an adult with one child (30%), an adult with two children (20%), two adults with one child (10%), and two adults with two children (5%).

Bartland's official Chief Economist has estimated that this subsidy program would tend to increase housing prices in a very tight housing market by an average of 0.001%. The Bartland City Council has $1000K$ available to reappropriate from elsewhere in the budget and would like to find the *best* way to use this budget to help with the houselessness problem. Both programs require staff support – in particular 10% of a full time equivalent staff member to process paperwork, conduct visits, and other service related activities. There are seven staff members available to work on these activities.

Let's summarize the data for two programs. Let's focus on expected numbers of people served for each policy intervention.

TABLE 8.1 Policy Options for Addressing Houselessness

Per unit	Tiny Homes (H)	Rent Subsidy (R)
1 adult	90%	15%
1 adult, 1 child	10%	30%
1 adult, 2 children	0%	20%
2 adults	0%	20%
2 adults, 1 child	0%	10%
2 adults, 2 children	0%	5%
Expected children served	0.1	0.9
Expected adults served	1.0	1.35
Expected total people served	1.1	2.25
Cost per unit (K)	\$10	\$25
Staff support per unit	0.1	0.1

One group on the city council wants to serve as many people (both children and adults) as possible while keeping under the total budget limit.

The second group feels that adults are responsible for their own situation but wants to save as many children from houselessness as possible within budget limits.

As usual, start by thinking of the decision variables. In this case, let's define H to be number of tiny homes to be built and R to be the rental housing subsidies provided. Of course these should be non-negative variables. We could use integer variables or continuous variables.

Next, let's look at our constraints and formulate them in terms of the decision variables. We have two constraints. The first one for the budget is simply: $10 \cdot H + 25 \cdot R \leq 1000$. The second is to ensure we have sufficient staff support, $0.1 \cdot H + 0.1 \cdot R \leq 7$.

Now, let's think about our objectives. The first group wants to serve as many people as possible so the objective function is $\max 1.1 \cdot H + 2.25 \cdot R$.

Similarly, since the second group is focused on children, their objective function is

$$\max 0.1 \cdot H + 0.9 \cdot R.$$

Let's put this all together in a formulation.

$$\begin{split} \max \; & 1.1 \cdot H + 2.25 \cdot R \\ \max \; & 0.1 \cdot H + 0.9 \cdot R \\ \text{s.t.} \; & 10 \cdot H + 25 \cdot R \leq 1000 \\ & 0.1 \cdot H + 0.1 \cdot R \leq 7 \\ & H, \; R \in \{0, 1, 2, ...\} \end{split}$$

Alas, linear programming models and the Simplex method only allows for a single objective function. Let's start by solving from the perspective of the first group.

$$\begin{split} \max \; & 1.1 \cdot H + 2.25 \cdot R \\ \text{s.t.} \; & 10 \cdot H + 25 \cdot R \leq 1000 \\ & 0.1 \cdot H + 0.1 \cdot R \leq 7 \\ & H, \; R \in \{0, 1, 2, ...\} \end{split}$$

```
Home1Model <- MIPModel() |>
  # Avoid name space conflicts by using a prefix of V
  add_variable(VH, type = "integer", lb = 0) |>
```

```
  add_variable(VR, type = "integer",lb = 0)  |>
  set_objective(1.1*VH + 2.25*VR,"max")      |>
  add_constraint(10*VH + 25*VR <= 1000)      |>
  add_constraint(0.1*VH + 0.1*VR <= 7)

res_Home1 <- solve_model(Home1Model,
                         with_ROI(solver = "glpk"))

H  <- get_solution (res_Home1 , VH)
R  <- get_solution (res_Home1 , VR)

sum_Home1           <- cbind(res_Home1$objective_value,H, R)
colnames(sum_Home1) <- c("Obj. Func. Val.", "H", "R")
rownames(sum_Home1) <- "Group 1: Max People"
```

TABLE 8.2 Group 1's Ideal Solution

	Obj. Func. Val.	H	R
Group 1: Max People	100	50	20

Now, let's examine the second group's model that has an objective of maximizing the expected number of children served.

$$
\begin{aligned}
\max\ & 0.1 \cdot H + 0.9 \cdot R \\
\text{s.t. } & 10 \cdot H + 25 \cdot R \leq 1000 \\
& 0.1 \cdot H + 0.1 \cdot R \leq 7 \\
& H,\ R \in \{0, 1, 2, ...\}
\end{aligned}
$$

```
Home2Model <- set_objective(Home1Model,
                            0.1*VH + 0.9*VR,"max")
  res_Home2 <- solve_model(Home2Model,
                           with_ROI(solver = "glpk"))

H  <- get_solution (res_Home2 , VH)
R  <- get_solution (res_Home2 , VR)

sum_Home2           <- cbind(res_Home2$objective_value,H, R)
colnames(sum_Home2) <- c("Obj. Func. Val.", "H", "R")
rownames(sum_Home2) <- "Group 2: Max Children"
```

TABLE 8.3 Comparing Homelessness Solutions

	Obj. Func. Val.	H	R
Group 1: Max People	100	50	20
Group 2: Max Children	36	0	40

So which group has the *better* model? The objective function value for group 1's model is higher, but it is in different units (people served) versus group 2's model of children served.

Both group's have admirable objectives. We can view this as a case of goal programming. By definition, we know that these are the best values that can be achieved in terms of that objective function. Let's treat these optimal values as targets to strive for and measure the amount by which fail to achieve these targets. We'll define target $T_1 = 100$ and $T_2 = 36$.

In order to do this, we need to use deviational variables. These are like slack variables from the standard form of linear programs. Since the deviations can only be one sided in this case, we only need to have deviations in one direction. We will define d_1 as the deviation in goal 1 (Maximizing people served) and d_2 as the deviation in goal 2 (Maximizing children served).

Let's now create the modified formulation.

$$
\begin{aligned}
\min\ & d_1 + d_2 \\
\text{s.t. } & 10 \cdot H + 25 \cdot R \leq 1000 \\
& 0.1 \cdot H + 0.1 \cdot R \leq 7 \\
& 1.1 \cdot H + 2.25 \cdot R + d_1 = T_1 = 100 \\
& 0.1 \cdot H + 0.9 \cdot R + d_2 = T_2 = 36 \\
& H,\ R \in \{0, 1, 2, ...\} \\
& d_1,\ d_2 \geq 0
\end{aligned}
$$

TABLE 8.4 Solution by Minimizing Sum of Deviations

	Obj. Func. Val.	H	R	d1	d1%	d2	d2%
Min sum of deviations	10	0	40	10	0.1	0	0

The deviation variables have different units though. One way to accommodate this would be to minimize the sum of percentages missed.

$$\begin{aligned}
\min \; & \frac{d_1}{T_1} + \frac{d_2}{T_2} \\
\text{s.t.} \; & 10 \cdot H + 25 \cdot R \leq 1000 \\
& 0.1 \cdot H + 0.1 \cdot R \leq 7 \\
& 1.1 \cdot H + 2.25 \cdot R + d_1 = T_1 \\
& 0.1 \cdot H + 0.9 \cdot R + d_2 = T_2 \\
& H, \; R \; \in \{0, 1, 2, ...\} \\
& d_1, \; d_2 \geq 0
\end{aligned}$$

```
## Status: optimal
## Objective value: 0.1
```

TABLE 8.5 Solution by Minimizing Sum of Percentage Deviations

	Obj. Func. Val.	H	R	d1	d1%	d2	d2%
Min sum of deviation %s	0.1	0	40	10	0.1	0	0

Another approach is to minimize the maximum deviation. This is often abbreviated as a *minimax*. This is essentially the same as the expression, "a chain is only as strong as its weakest link". In Japan, there is an expression that the "the nail that sticks up, gets pounded down" and in China, "the tallest blade of grass gets cut down." We can implement the same idea here by introducing a new variable, Q that must be at least as large as the largest miss.

$$\begin{aligned}
\min \; & Q \\
\text{s.t.} \; & 10 \cdot H + 25 \cdot R \leq 1000 \\
& 0.1 \cdot H + 0.1 \cdot R \leq 7 \\
& 1.1 \cdot H + 2.25 \cdot R + d_1 = T_1 \\
& 0.1 \cdot H + 0.9 \cdot R + d_2 = T_2 \\
& Q \geq \frac{d_1}{T_1} \\
& Q \geq \frac{d_2}{T_2} \\
& H, \; R \; \in \{0, 1, 2, ...\} \\
& d_1, \; d_2 \geq 0
\end{aligned}$$

Let's show the full R implementation of our minimax model.

```
Home5Model <- MIPModel() |>
 # To avoid name space conflicts, using a prefix of V
 #    for ompr variables.
  add_variable(VQ, type = "continuous")             |>
  add_variable(VH, type = "integer", lb = 0)        |>
  add_variable(VR, type = "integer",lb = 0)         |>
  add_variable(Vd1, type = "continuous",lb = 0) |>
  add_variable(Vd2, type = "continuous",lb = 0) |>
  set_objective(VQ ,"min")                          |>
  add_constraint(VQ>=Vd1/T1)                        |>
  add_constraint(VQ>=Vd2/T2)                        |>
  add_constraint(1.1*VH + 2.25*VR + Vd1 == T1)  |>
  add_constraint(0.1*VH + 0.9*VR + Vd2 == T2)   |>
  add_constraint(10*VH + 25*VR <= 1000)             |>
  add_constraint(0.1*VH + 0.1*VR <= 7)

res_Home5 <- solve_model(Home5Model,
                         with_ROI(solver = "glpk"))
res_Home5
```

```
## Status: optimal
## Objective value: 0.08
```

```
H  <- get_solution (res_Home5, VH)
R  <- get_solution (res_Home5, VR)
d1 <- get_solution (res_Home5, Vd1)
d2 <- get_solution (res_Home5, Vd2)

sum_Home5  <- cbind(
  res_Home5$objective_value, H, R,
  d1, d1/T1, d2, d2/T2)
colnames(sum_Home5) <-
  c("Obj. Func. Val.", "H", "R", "d1", "d1%", "d2", "d2%")
rownames(sum_Home5) <- "Minimax"
```

The minimax solution finds an alternative that is still Pareto optimal.

Careful readers may note that children are effectively double counted between the two objective functions when deviations are added.

This example can be expanded much further in the future with additional policy interventions, other stakeholders, and other characteristics, such as policies

TABLE 8.6 Minimax Solution

	Obj. Func. Val.	H	R	d1	d1%	d2	d2%
Min sum of deviations	10.00	0	40	10	0.10	0.0	0.0000000
Min sum of deviation %s	0.10	0	40	10	0.10	0.0	0.0000000
Minimax	0.08	10	36	8	0.08	2.6	0.0722222

on drug addiction treatment, policing practices, and more. We did not factor in the Chief Economist's impact on housing prices. We'll leave these issues to future work.

8.4 Mass Mailings

Let's take a look at another example. We have a mailing outreach campaign across the fifty states to do in the next eight weeks. You have C_s customers in each state. Since the statewide campaign needs to be coordinated, each state should be done in a single week but different states can be done in different weeks. You want to create a model to have the workload, in terms of numbers of customers, to be as balanced as possible across the eight weeks.

As an exercise, pause to think of how you would set this up.

What are your decision variables?

What are your constraints?

What is the objective function?

Try to give some thoughts as to how to set this up before moving on to seeing our formulation.

To provide some space before we discuss the formulation, let's show the data. Rather than providing a data table that must be retyped, let's use a dataset already available in R so you can simply load the state data. Note that you can grab the population in 1977 in terms of thousands.

```
data(state)
Customers <- state.x77[,1]
kbl (head (Customers), booktabs=T,
     caption="Number of Customers for First Six States.") |>
  kable_styling(latex_options = "hold_position")
```

TABLE 8.7 Number of Customers for First Six States

	x
Alabama	3615
Alaska	365
Arizona	2212
Arkansas	2110
California	21198
Colorado	2541

8.4.1 Formulating the State Mailing Model

Presumably you have created your own formulation. If so, your model will likely differ from what follows in some ways such as naming conventions for variables or subscripts. That is fine. The process of trying to build a model is important.

Let's start by defining our decision variables, $x_{s,w}$ as a binary variable to indicate whether we are going to send a mailing to state s in week w.

Now, we need to ensure that every state is mailed to in one of the eight weeks. We simply need to add up the variable for each state's decision to mail in week 1, 2, 3, ..., up to 8. Mathematically, this would be $\sum\limits_{w=1}^{8} x_{s,w} = 1, \ \forall \ s$.

It is useful to take a moment to reflect on why $\sum\limits_{s=1}^{50} \sum\limits_{w=1}^{8} x_{s,w} = 50$ is not sufficient to ensure that all 50 states get mailed to during the eight week planning period.

Combined with the variable $x_{s,w}$ being defined to be binary, this is sufficient to ensure that we have a *feasible* answer but not necessarily a well-balanced solution across the eight weeks.

We could easily calculate the amount of material to mail each as a function of x_s, w and C_s. For week 1, it would be $\sum_{s=1}^{50} C_s \cdot x_{s,1}$ For week 2, it would be $\sum_{s=1}^{50} C_s \cdot x_{s,2}$, and so on. This could be generalized as $\sum_{s=1}^{50} C_s \cdot x_{s,w} \; \forall \; w$.

Creating a balanced schedule can be done in multiple ways. Let's start by using the *minimax* approach discussed earlier. To do this, we add a new variable Q and constrain it to be at least as large as each week. Therefore, for week 1, $Q > \sum_{s=1}^{50} C_s \cdot x_{s,1}$ and for week 2, $Q \geq \sum_{s=1}^{50} C_s \cdot x_{s,2}$. Again, to generalize for all eight weeks, we could write $Q \geq \sum_{s=1}^{50} C_s \cdot x_{s,w} \; \forall \; w$.

we can then use our minimax objective function of simply minimizing Q.

We'll summarize our formulation now.

$$
\begin{aligned}
&\min Q \\
&\text{s.t.} \sum_{w=1}^{8} x_{s,w} = 1, \; \forall \; s \\
&\quad Q \geq \sum_{s=1}^{50} C_s \cdot x_{s,w} \; \forall \; w \\
&\quad x_{s,w} \; \in \{0,1\} \; \forall \; s, \; w
\end{aligned}
$$

8.4.2 Implementing the State Mailing Model

Let's now move on to implement this model in R.

```
States <- 50  # Options to shrink problem for testing
Weeks <- 8

Mail1 <- MIPModel()
      # 1 iff state s gets assigned to week w
Mail1 <- add_variable(Mail1, Vx[s, w],
                      s=1:States, w=1:Weeks, type="binary")
```

```
Mail1 <- add_variable(Mail1, VQ, type = "continuous")

Mail1 <- set_objective(Mail1, VQ, "min")

# every state  needs to be assigned to a week
Mail1 <- add_constraint(Mail1, sum_expr(Vx[s, w],
                        w=1:Weeks)==1, s=1:States)
Mail1 <- add_constraint(Mail1, VQ >=
                        sum_expr(Customers [s]*Vx[s, w],
                                 s=1:States), w = 1:Weeks)
Mail1
```

```
## Mixed integer linear optimization problem
## Variables:
##   Continuous: 1
##   Integer: 0
##   Binary: 400
## Model sense: minimize
## Constraints: 58
```

```
res_Mail1 <- solve_model(Mail1,
                         with_ROI(solver = "symphony",
                  verbosity=-1, gap_limit=1.5))
```

```
res_Mail1
```

```
## Status: infeasible
## Objective value: 26834
```

Note that the messages from Symphony indicate that the solution found was feasible while `ompr` interprets the status as `infeasible`. This is a bug that we have discussed earlier. Turning on an option for more messages from the solver such as `verbose=TRUE` for `verbose=TRUE` or `verbosity=0` for `symphony` can give confirmation that the final status is *not* infeasible.

Another useful to thing to note is that solving this problem to optimality can take a long time despite having fewer binary variables than some of our earlier examples. Using `glpk` with `verbose=FALSE` means that the MIP is solved with no progress information displayed and makes it look like the solver is hung.

Turning on more information (increasing the verbosity) helps explain that the Solver is working, it is just taking a while, on my computer I let it run 20 minutes without making further progress than a feasible solution it had found quickly.

In fact, I realized that the feasible solution found was very close to the best remaining branches so perhaps this solution was optimal, but it was taking a very long time to prove that it was optimal. In any case, it is probably good enough. Often data may only be accurate to $\pm5\%$ so spending extensive time trying to get significantly more accurate results is not very productive. This suggests setting stopping options such as a time limit, number of LPs, or a *good enough* setting. For this problem, I chose the latter option.

We solved this with a mixed integer programming problem gap limit of 1.5% meaning that while we have not *proven* this solution to be optimal, we do know that it is impossible to find a solution more than 1.5% better. From a branch and bound algorithm perspective, this means that while we have not searched down fully or pruned every branch, we know that no branch has the potential of being more than 1.5% better than the feasible solution that we have already found.

Now let's move on to discussing the results. We will start with filtering out all the variables that have zero values so we can focus on the ones of interest – the states that are assigned to each week. Also, notice that a `dplyr` function was used to add state names to the data frame.

```
assigned1a <- res_Mail1 |>
 get_solution(Vx[s,w]) |>
 filter(value >.9) |>
 select (s,w)
rownames(assigned1a)<-c(names(Customers[assigned1a[,1]]))

kbl (head(assigned1a), booktabs=T,
     caption="Example of some states assigned to week 1") |>
  kable_styling(latex_options = "hold_position")
```

That is just for six states in the first week though.

```
assigned1b <- tibble::rownames_to_column(assigned1a)
table_results1<-c() # Prepare a new table of results
cols_list<-c()      # Prepare column list
weeksmail<-""
```

TABLE 8.8 Example of Some States Assigned to Week 1

	s	w
Indiana	14	1
Maryland	20	1
Minnesota	23	1
Nevada	28	1
North Carolina	33	1
South Carolina	40	1

```
for (week_counter in 1:Weeks) {
  weeksmail1a<- assigned1b |>
  filter(w==week_counter) |>
  select(rowname)
  cols_list <- append(cols_list, paste0("Week ", week_counter))
  table_results1 <- append(table_results1,
                           paste(weeksmail1a$rowname,
                                 collapse = ", "))}

text_tbl <- data.frame(Weeks = cols_list,States = table_results1)
```

```
kbl(text_tbl, booktabs=T,
     caption="Display of States by Week") |>
  kable_styling(latex_options = c("hold_position")) |>
  column_spec(1, bold = T) %>%
  column_spec(2, width = "30em")
```

Since this is state level data, let's look at a map of the schedule.

```
library (ggplot2)
library (maps)
mapx = data.frame(region=tolower(rownames(assigned1a)),
    week=assigned1a[,"w"],
    stringsAsFactors=F)

states_map <- map_data("state")
```

TABLE 8.9 Display of States by Week

Weeks	States
Week 1	Indiana, Maryland, Minnesota, Nevada, North Carolina, South Carolina, Tennessee
Week 2	Alabama, Arkansas, Delaware, Iowa, Montana, South Dakota, Texas, Washington
Week 3	Alaska, Massachusetts, Missouri, Oregon, Pennsylvania, Utah, Vermont
Week 4	Arizona, Hawaii, Idaho, Illinois, Kentucky, Maine, New Hampshire, West Virginia, Wisconsin
Week 5	Colorado, Georgia, Ohio, Oklahoma, Rhode Island, Virginia
Week 6	Florida, New York, Wyoming
Week 7	California, Louisiana, New Mexico
Week 8	Connecticut, Kansas, Michigan, Mississippi, Nebraska, New Jersey, North Dakota

```
ggplot(mapx, aes(map_id = region)) +
    geom_map(aes(fill = week, colour = "white"),
             map = states_map) +
scale_fill_viridis_c(option = "C") +
expand_limits(x = states_map$long, y = states_map$lat)
```

Note that this leaves off Alaska and Hawaii for visualization. For completeness, Alaska is in week 3 and Hawaii is in week 4.

8.4.3 Frontloading the Work

This formulation generates solutions that have high symmetry. Essentially it would be feasible and have the same objective function value if we simply swap any two weeks of assignments. For example, if we swap all of the states assigned to week 1 and week 2, it would still be feasible and have exactly the same objective function value. This would result in a high number of alternate optimal solutions. The numbering by week is essentially arbitrary since the weeks don't make a difference.

When there exists a high degree of symmetry, it may be useful to implement one or more constraints to *break* the symmetry by differentiating between solutions. One approach to doing this is to require a particular ordering of weeks. For example, we could require that weeks get progressively lighter in terms of workload. In fact, this might be a managerial preference to ensure that if problems develop, there is some organizational slack later to deal with this.

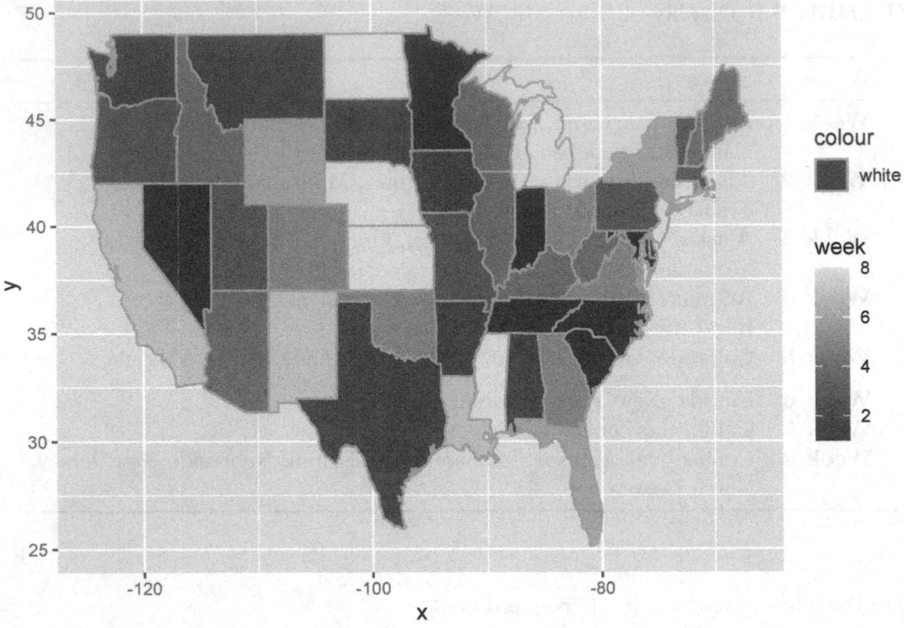

FIGURE 8.1 Map of Mailing Results.

$$\sum_{s=1}^{50} C_s \cdot x_{s,w-1} \geq \sum_{s=1}^{50} C_s \cdot x_{s,w}, \; w \in \{2, \ldots, 8\}$$

We could then extend our `ompr` model, `Mail1` with the additional constraints.

```
Mail2 <- add_constraint(Mail1,
                    sum_expr(Customers [s]*Vx[s, w-1],
                              s=1:States) >=
          sum_expr(Customers [s]*Vx[s, w],
                              s=1:States),
          w = 2:Weeks)
```

While it might be thought that adding constraints limits the search space of possible solutions and might speed up solution, in this case it slows down solving speed tremendously. Despite having having less than half the binary variables of applications from Chapter 8 and the same number the earlier mail problem, this problem turns out to be most computationally demanding so far. The branch and bound process from symphony takes hours to solve.

```
kbl(as.matrix(table_results1[-1]), booktabs=T, caption=
      "Solution with Symmetry Breaking") |>
  kable_styling(latex_options = "hold_position")
```

The solving time using this modification to the LP changed tremendously. On an Intel i7 mobile CPU and R 4.1.1. the following results were obtained before terminating the solver after nearly four hours. Within 6 seconds it found a feasible solution as demonstrated by the Lower Bound of 26,540 but the upper bound of what could still be undiscovered was 29,602. This gives a possible suboptimality gap of 10.34%. Over the next half hour the Solver did not find a better solution (Lower Bound stays unchanged) but is able to reduce the Upper Bound. At this point I terminated the solver. There had been modest improvement in the search process in the first 30 minutes with two and a half million iterations (linear programming subproblems) but then no change for the three and half hours and almost six million more iterations.

TABLE 8.10 Solving Progress for Modified Mailing Model

Seconds	Iterations	Lower Bound	Upper Bound	Gap (%)
6	9,397	26540.12	29602.00	10.34
1291	2,015,431	26540.12	**28940.00**	8.24
1711	2,508,415	26540.12	**28510.00**	6.91
14311	8,391,692	26540.12	28510.00	6.91

This highlights that large or complex optimization problems may take extra effort and attention. Analysts can make judicious tradeoffs of potentially suboptimal results, solving time, and data accuracy. For our purpose, to speed up the solution time, you could do an early termination option such as setting the `gap_limit` to 15% by adding `gap_limit=15`. Setting an appropriate termination parameter based on gap, number of iterations, or time can be particularly helpful while developing a model. It also underscores that the benefits of implementing and testing algebraic models with smaller sized problems before expanding to the full sized problems. By separating the data from the model, we can easily just shrink or expand the problem without affecting the formulation or the implementation. In this case, we could shrink the problem by using fewer weeks and states by changing the upper limits (`Weeks<-10; States<-10`) used in the model, resulting in 30 binary variables instead of 400.

We could further modify the model to eliminate Q since the workload in week 1 would essentially be the heaviest workload (tallest nail.) This highlights that in optimization and particularly integer programming, there are often many different ways to implement models for the same application. This underscores the importance of clearly defining the elements of the formulation.

This application and model can be adjusted to fit a wide variety of other situations such as:

- Setting a maximum or minimum number of states per week
- Having a maximum and/or minimum number of customers to mail per week.
- Incorporating a secondary goal of finishing the mailing in as few weeks as possible.
- Applying this approach to other applications such as assigning customers to salespeople.

8.5 Exercises

Exercise 8.1 (Adjustments to Houselessness Policies). From the houselessness problem given in chapter, make changes as given below: First option is to build new, government subsidized tiny homes for annual cost of $15K which would serve one adult 85% of the time and a parent with a child 15% of the time. Another option is to create a rental subsidy program which costs $35K per year per unit which typically serves a single adult (15%), two adults (15%), an adult with one child (35%), an adult with two children (20%), two adults with one child (10%), and two adults with two children (5%).

Solve the problem and Discuss how the solution has changed.

Exercise 8.2 (Adjustments to Houselessness Policies). Create and justify your own third policy option to enhance the original exercise in the book. Describe the policy in the context of the application. Formulate, implement, and solve the revised version.

Does the revised solution match your expected result?

Exercise 8.3 (Extending State Mailing). The mailing by state example in chapter 8 is just another example of mixed integer programming.

a. Write the optimization model using LaTeX to ensure that no week has more than 8 states assigned to the week.

b. Write a constraint or set of constraints using LaTeX that would ensure state 3 is done before state 7.

c. Write a constraint or set of constraints to ensure that state 9 and 24 are done in the same week.

d. Given the minimax objective function, how do you think the above requirements will affect the solution in terms of an objective function and characteristics of managerial interest?

Exercise 8.4 (Experimenting with State Mailing). In the mailing by state example in chapter, there is a change discovered that Week 4 is going to be during a Holiday week and the Mailing Center will be closed for whole week. Make appropriate changes to the model in example presented in the chapter to reflect these changes.

a. Write the optimization model using LaTeX to ensure that no week has more than 8 states assigned to the week.
b. Write a constraint or set of constraints using LaTeX that would ensure state 4 is done before state 6.
c. Write a constraint or set of constraints to ensure that state 5 and 9 are done in the same week.
d. Write a constraint or set of constraints to ensure that the states of Oregon and California are done before holiday week.
e. After solving above, compare the results with exercise 8.1 and discuss.

A

A Very Brief Introduction to R

A.1 Purpose

This Appendix provides a very brief introduction to the basics of R from the perspective of use for optimization modeling. This is not meant to be comprehensive R tutorial. There is an ever expanding collection of such materials available.

A.2 Getting Started with R

This Appendix helps a reader new to R quickly get started. I strongly recommend using RStudio. I also like books such as *R in a Nutshell* (Adler, 2012) or *R in Action* (Kabacoff, 2011).

I usually find the best way for me to learn a new tool is to roll up my sleeves and jump right in. This book can be approached in that way – just be ready for a little more experimentation. This book is available on Github and the R Markdown files are available for use. Code fragments are shown quite liberally for demonstrating how things work. While this may be a bit verbose it is meant to enable readers to jump in at various points of the book.

While code fragments can be copied from the R markdown files for this book, it is often best to physically retype many of the code fragments shown as that gives time to reflect on what each statement is doing.

Let's define some conventions to be used throughout the book. First, let me be clear, the goal of this section is not to provide a comprehensive introduction to R. Entire books are written on that subject. My goal here is to simply make sure that everyone can get started productively in the material covered later.

Since this book is focused on using R for operations research, we will focus on the capabilities that are needed for this area and introduce additional features

DOI: 10.1201/9781003051251-A

and functions as needed. If you are already comfortable with R, RStudio, and RMarkdown, you may skip the remainder of this section.

Begin by ensuring that you have access to or installed R and RStudio. Both are available for Windows, Mac, and Linux operating systems as well as being available as web services.

Now, let's assume that you are running RStudio.

In this book, I will frequently show code fragments called chunks to show R code. In general, these code chunks can be run by simply retyping the command(s) in the Console.

```
a <- 7
```

This command assigns the value of 7 to the variable *a*. It is standard in R to use `<-` as an assignment operator rather than an equals sign. We can then use R to process this information.

```
6*a
```

```
## [1] 42
```

Yes, not surprisingly, $6 * 7$ is 42. Notice that if we don't assign that result to something, it gives an immediate result to the screen.

We will often be using or defining data that has more than one element. In fact, R is designed around more complex data structures. Let's go ahead and define a matrix of data.

```
b<-matrix(c(1,2,3,4,5,6,7,8))
```

By default, it is assuming that the matrix has one column which means that every data value is in a separate row. The `c` function is a concatenate operator meaning that it combines all the following items together.

Let's look at *b* now to see what it contains.

```
b
```

```
##      [,1]
## [1,]    1
## [2,]    2
## [3,]    3
## [4,]    4
## [5,]    5
## [6,]    6
## [7,]    7
## [8,]    8
```

Let's instead define this matrix to have four columns and two rows.

```
b<-matrix(c(1,2,3,4,5,6,7,8), ncol=4)
b
```

```
##      [,1] [,2] [,3] [,4]
## [1,]    1    3    5    7
## [2,]    2    4    6    8
```

Notice that since there are eight elements, we only needed to tell R that the matrix has four columns, and it then knew that there would be only two rows. Of course we could have set the number of rows to be two for the same result.

This is still a little ambiguous. Let's give the rows and columns names. For now we will simply name them Row and Col.

```
b<-matrix(c(1,2,3,4,5,6,7,8), ncol=4,
          dimnames=c(list(c("Row1", "Row2")),
                     list(c("Col1", "Col2","Col3","Col4"))))
```

Remark (RStudio Console). The RStudio console can save typing by pressing the up arrow key to view previous command(s) which can then be further edited.

Okay, this command has a lot more going on. The term `dimnames` is a parameter that contain names for rows and columns. One thing to note is that this line fills up more space than a single line, so it rolls over to multiple line. The `dimnames` parameter will get get two concatenated (combined) lists. The first list is a combined list of two text strings "Row1" and "Row2". The next line does the same for columns.

Let's confirm that it works.

```
b
```

```
##      Col1 Col2 Col3 Col4
## Row1    1    3    5    7
## Row2    2    4    6    8
```

This table is still not that "nice" looking. Let's use a package that does a nicer job of formatting tables. To do this, we will use an extra package. Up to this point, everything that we have done just simply uses standard built in functions of R. The package that we will use is `kable` but there are plenty of others available such as `pander`, `kable`, `xtable`, and `huxtable`. While `kable` is a function of `knitr`, it is significantly enhanced by the `kableExtra` package. For more information on the specifics of creating rich tables, see Appendix D.

Let's start by loading the `kableExtra` package.

```
library (kableExtra)
```

If R indicates that you don't have the `kableExtra` package installed on your computer, you can press the "Install" command under the Packages tab and then type in `kableExtra` or use the `install.packages` command from RStudio. The `kbl` is a shorthand way of entering the `kable` function that is provided by `kableExtra`. A lot of the power of `R` comes from the thousdands of packages developed by other people so you will often be installing and loading packages.

Notice the hash symbol is used to mark a comment in the above code chunk. It is also used to "comment out" a command that I don't need to use at the current time. Using comments to explain what a command is doing can be helpful to anyone that needs to revisit your code in the future, including yourself!

```
knitr::kable (b, booktabs=T,
    caption="Example using Kable with booktabs") |>
  kableExtra::kable_styling(latex_options = "hold_position")
```

The table looks very nice in the book. Let's explain in detail how the table is generated. This code chunk has a lot going and is worth a careful look.

TABLE A.1 Example Using kable with booktabs

	Col1	Col2	Col3	Col4
Row1	1	3	5	7
Row2	2	4	6	8

- The first line `knitr::kable` means that we should run the `kable` function from within `knitr`. In general, as long as the library is loaded and the same function is not defined by two different loaded libraries, you don't need to specify where the function is being run from.
- The `booktabs=T` sets a format for clean published table style by setting the option to `T` for `TRUE`.
- The second line provides a caption.
- The last part of the second line warrants a little attention. It is the new pipe operator and requires `R` version 4.1.0 or higher. The pipe operator basically says pass this commands output into the beginning of the next command. We'll use it a lot for table generation. Prior to R version 4.1.0, people that used pipes in R often used the `%>%` operator from a separate package such as `magrittr`. For most users, including us, the built-in `|>` will suffice.
- The last command, `kable_styling`, has an option for `hold_position` which holds the placement closer to where it is called, keeping LaTeX from second guessing where the ideal location would be for the table. Note that is also specifying its source package, in this case, `kableExtra`. There are a lot of other options in `kable_styling` that we will use in later chapters as needed.
- For your own use, you could shorten all of this to just `kable (b)`. On the other hand, `kableExtra` has one more small trick up its sleeve. It provides a shorter name for the `kable` function of just `kbl`. When space is at premium, this shortcut is handy. It also serves as a helpful check to ensure that `kableExtra` is loaded.

Let's continue experimenting with operators.

```
c<-a*b
```

TABLE A.2 Scalar Multiplication of Matrix b to Make Matrix c

	Col1	Col2	Col3	Col4
Row1	7	21	35	49
Row2	14	28	42	56

Now let's do a transpose operation on the original matrix. What this means that it converts rows into columns and columns into rows.

```
d<-t(b)
```

TABLE A.3 Transposition of Matrix b

	Row1	Row2
Col1	1	2
Col2	3	4
Col3	5	6
Col4	7	8

Note:
Row and column names were changed at the same time.

Notice that row and column names are also transposed. The result is that we now have rows labeled as columns and columns labeled as rows! This is only a problem given that we used the words rows and columns in the names. If these had been more descriptive such as weeks and product names, it would have been a good thing to change them at the same time.

Now we have done some basic operations within R.

We could try this, but we aren't quite there yet.

```
c
```

```
##      Col1 Col2 Col3 Col4
## Row1    7   21   35   49
## Row2   14   28   42   56
```

```
d
```

```
##      Row1 Row2
## Col1    1    2
## Col2    3    4
## Col3    5    6
## Col4    7    8
```

Try renaming the rows and columns for `d`. As a hint, you could use `dimnames` or `colnames` and `rownames`.

Recall that we have created a variety of objects now: a, b, c, and d. R provides a lot of tools for slicing, dicing, and combining data.

TABLE A.4 Original Matrix b

	Col1	Col2	Col3	Col4
Row1	1	3	5	7
Row2	2	4	6	8

Let's look at the original matrix, b and what we can do with it. Let's grab the second row, third column and last element.

TABLE A.5 Second Row of b

Col1	2
Col2	4
Col3	6
Col4	8

The `as.matrix` function is used to convert the object into a matrix so that `kable` can display it well.

```
temp2 <- as.matrix(b[,3])
```

TABLE A.6 Third Column of b

Row1	5
Row2	6

Grabbing the third column is not terribly interesting but operations such as this will often be quite useful.

```
temp3 <- as.matrix(b[2,4])
```

TABLE A.7 Last Element of b

8

Let's take the first row of `b` and combine it with the second row of `c` to form a new matrix, `e`. Since these are rows that are going to be combined, we will use a command, `rbind`, to bind these rows together.

```
temp4 <- rbind(b[1,],c[2,])
```

TABLE A.8 Combined Matrix

Col1	Col2	Col3	Col4
1	3	5	7
14	28	42	56

Notice that *temp4* has inherited the column names but lost the row names. Let's set the row names for this matrix.

```
rownames(temp4)<-list("From b", "From c")
```

TABLE A.9 Combined Matrix with Explanation of Source

	Col1	Col2	Col3	Col4
From b	1	3	5	7
From c	14	28	42	56

We could combine all of matrix `b` and matrix `c` together using row binding or column binding. Table Let's view the results of binding using columns (`cbind`) and rows (`rbind`).

```
temp5<- cbind(b,c)
```

TABLE A.10 Column Binding of Matrices b and c

	Col1	Col2	Col3	Col4	Col1	Col2	Col3	Col4
Row1	1	3	5	7	7	21	35	49
Row2	2	4	6	8	14	28	42	56

```
temp6 <- rbind(b,c)
```

TABLE A.11 Row Binding of Matrices b and c

	Col1	Col2	Col3	Col4
Row1	1	3	5	7
Row2	2	4	6	8
Row1	7	21	35	49
Row2	14	28	42	56

Data organizing is a less glamorous part of the job for practicing analytics professionals but can consume a majority of the workday. There is a lot more that can be said about data wrestling but scripting the data cleansing in terms R commands will make the work more repeatable, consistent, and in the end save time.

A.3 Exercises

Exercise A.1 (Creating a Matrix of Daily Demand for Four Weeks)**.** Assume that your company's product has a demand of 10 widgets on Monday, increasing by five units for every day through Sunday. Build a matrix of four weeks of demand where each row is a separate week. Each Monday starts over with the same demand. The rows should be named Week1, Week2, etc. The columns should have names corresponding to the day of the week.

Exercise A.2 (Creating a Matrix of Demand for Two Products)**.** Assume that your company's product has a demand of 10 widgets on Monday, increasing by five units for every day through Sunday. Gadgets have a demand of 20 on Monday, increasing by 3 units a day through Sunday. Build a matrix showing each product as a separate row. Rows should have names for the products and columns for the days of the week.

Exercise A.3 (Creating a Matrix of Monthly Demand)**.** Consider some top selling online products for the year 2021: toys, shoes, pens and pencils, decorative bottles, drills, cutters, and GPS navigation systems. Create your own assumed starting counts per item for throughout year by increasing each item sales by 50 units per month, starting from January till December.

Exercise A.4 (Displaying Selective Data from the Matrix)**.** In the above matrix, grab the columns and rows to show the data only for pens and pencils during the months of June, August and October.

Exercise A.5 (Creating Separate Demands for Products). In the above matrix, demand for toys and shoes has been increased by 20 and 10 units, respectively. Create a matrix for only these products for the first 6 months. Display the matrix twice–once without formatting and a second time using an R package that provides better table formatting (`pander`, `kable` via `knitr`, `huxtable` or similar tool. Be sure to provide a table caption.)

B

Introduction to Math Notation

B.1 Purpose

This Appendix provides a brief introduction to relevant notation constructs that arise often in optimization. In this appendix we will both briefly review some of these mathematical relationships that commonly arise in mathematical models as well as how these can be expressed using LaTeX in Rmarkdown. A major benefit of using the RMarkdown workflow is that it enables precise mathematical descriptions alongside the actual analysis.

B.2 Basic Summation Notation

In linear programming, we often need to add together a lot of numbers (or variables), such as: $1 + 2 + 3 + 4 + 5 + 6 + 7 + 8 + 9 + 10 + 11 + 12 + 13 + 14 + 15 + 16 + 17 + 18 + 19 + 20 + 21 + 22 + 23 + 2 + 25 + 26 + 27 + 28 + 29 + 30 + 31 + 32 + 33 + 34 + 35 + 36 + 37 + 38 + 39 + 40 + 41 + 42 + 43 + 44 + 45 + 46 + 47 + 48 + 49 + 50$

Summation notation is a mathematical shorthand used to represent repeated addition of algebraic expressions involving one or more variables. The Greek capital letter sigma, $\sum$, is the symbol used to show that we wish to make calculations using summation notation. Since summation notation includes at least one variable – let's limit this initial examination to one variable only – the variable is displayed below the $\sum$ symbol. The summation notation below tells us that we will be finding the summation of a variable expression involving x.

$$\sum_{x}$$

The starting value of the variable may be given below $\sum$ as well. In this case,

DOI: 10.1201/9781003051251-B

the summation notation below tells us that the initial value of x will be equal to 1.

$$\sum_{x=1}$$

In some cases, we may also wish to designate an ending value of the variable, which we can include above the $\sum$. The summation notation below also tells us that the final value of x will be equal to 5.

$$\sum_{x=1}^{5}$$

In all dealings with summation notation, variables will only take integer values, beginning and ending at any values provided within the summation notation. Note that some summations may use an "ending" value of ∞, which would involve the summation of an infinite number of values.

Let's look at a basic summation problem.

$$\sum_{x=1}^{5} 2x$$

The summation above means that we will take the x values, starting at $x = 1$, and multiply the value of x by 2. We will continue to do this for each integer value of x until we have reached our ending value of 5. Then we will sum all of our results (five of them, in this case) to produce one final value for the summation.

$$\begin{aligned}\sum_{x=1}^{5} 2x &= 2 \cdot 1 + 2 \cdot 2 + 2 \cdot 3 + 2 \cdot 4 + 2 \cdot 5 \\ &= 2 + 4 + 6 + 8 + 10 \\ &= 30\end{aligned}$$

Summation can be calculated over a variety of algebraic expressions.Another, perhaps more challenging, example is shown below.

$$\begin{aligned}&\sum_{x=0}^{3} x^2 - 4x + 1 \\ &= (0^2 - 4 \cdot 0 + 1) + (1^2 - 4 \cdot 1 + 1) + (2^2 - 4 \cdot 2 + 1) + (3^2 - 4 \cdot 3 + 1) \\ &= 1 + (-2) + (-3) + (-2) \\ &= -6\end{aligned}$$

B.3 Using LaTeX in RMarkdown

LaTex is a common typesetting used to express mathematical notation. The summation notation symbols thus far in this text have been written using LaTeX. Inline LaTeX commands can be added by including a $ symbol on either side of the LaTeX command. To create entire LaTeX code chunks, include two $$ symbols before and after the chunk of LaTeX.

B.4 Inline Notation

Rmarkdown makes it easy to switch between regular and "inline LaTeX" by selecting whether you use a single or a double $ to wrap the math. Most of the above examples were written using the double $.

Inline tries to help fit the math in the normal height of a line of text. For example, notice how this equation doesn't really disrupt the line spacing in a paragraph $\sum_{x=0}^{3} x^2 - 4x + 1$. The most obvious difference is in the summation. If it is selected as in-line, the summation limits will be to the right of the summation symbol. If it is regular, the limits will be above and below the summation symbol. As an example, here is the same equation

$$\sum_{x=0}^{3} x^2 - 4x + 1$$

entered in the middle with regular text. Entering a regular summation in the paragraph causes problems.

Let's try the opposite now by showing what an in-line summation looks like when entered on its own line.

$\sum_{x=0}^{3} x^2 - 4x + 1$

Again, in this case, the summation is not quite right. Given that vertical space is no longer at a premium, it is just harder to read than the regular summation.

The simple rule of thumb in Rmarkdown is to use a double-dollar sign for equations that are entered on their own line and a single dollar sign for symbols that are entered as text.

B.5 Sums

To display the $\sum$ symbol in LaTex, use the command `\sum_{lower}^{upper}`. The lower and upper limits of summation after the `\sum` are both optional. The upper and sometimes the lower are omitted when the interpretation would be unambiguous. The summation expression can be added using the command:

$$\sum_{lower}^{upper}$$

Sum limits can be written to appear above and below the operator using the same notation as for superscript and subscript.

$$\sum_{t=0}^{n} \frac{CF_t}{(1+r)^t}$$

B.6 Delimiters

Delimiters, like parentheses or braces, can automatically re-size to match what they are surrounding. This is done by using `\left` and `\right` before the particular parentheses or brace being used. This is a well formatted example:

$$\left(\sum_{i=1}^{n} i\right)^2 = \left(\frac{n(n-1)}{2}\right)^2 = \frac{n^2(n-1)^2}{4}$$

For illustration purposes, let's at the code for the right-hand side.

```
\left( \frac{n(n-1)}{2}\right)^2 = \frac{n^2(n-1)^2}{4}
```

If we had not used the `\left` and `\right` the original equation would have looked like this.

$$(\sum_{i=1}^{n} i)^2 = (\frac{n(n-1)}{2})^2 = \frac{n^2(n-1)^2}{4}$$

B.7 Summary of Mathematical Notations

Below are some common mathematical functions that are often used.

TABLE B.1 Commonly Used Optimization Modeling Notations

Math	LaTeX	Purpose
$x = y$	`x = y`	Equality
$x < y$	`x < y`	Strict less than
$x > y$	`x > y`	Strict greater than
$x \le y$	`x \le y`	Strict less than
$x \ge y$	`x \ge y`	Strict greater than
$x \ne y$	`x \ne y`	Not equal
x^{n}	`x^{n}`	Superscript or exponentiation
$x^{\text{Max Demand}}$	`x^{\text{Max Demand}}`	Superscript
x_i	`x_i`	Subscript
$x_{i,j,k}$	`x_\{i,j,k\}`	Subscript
$\overline{x}$	`\overline{x}`	Bar
$\hat{x}$	`\hat{x}`	Hat
$\tilde{x}$	`\tilde{x}`	Tilde
$\frac{a}{b}$	`\frac{a}{b}`	Fraction or Ratio
$1 + 2 + \cdots + 10$	`1 + 2 + \cdots + 10`	Ellipsis
$\lvert A \rvert$	`|A|`	Absolute Value
$x \in A$	`x \in A`	x is in set A
$x \subset B$	`x \subset B`	x is a proper subset of B
$x \subseteq B$	`x \subseteq B`	x is a subset of B
$A \cup B$	`A \cup B`	Union of sets A and B
$1, 2, 3$	`\{1, 2, 3\}`	Set with members 1, 2, and 3
$\int_{a}^{b} f(x) \; dx$	`\int_{a}^{b} f(x) \; dx`	Integral
$\sum_{x=a}^{b} f(x)$	`\sum_{x = a}^{b} f(x)`	Summation
λ	`\lambda`	Greek letter or symbol
$3 \cdot x$	`3 \cdot x`	Use dot for multiplication

Of course there are many more mathematical terms, Greek letters, and more but this table can serve as a common quick reference for things that may come up in frequently in optimization modeling.

B.8 Sequences and Summation Notation

Often, especially in the context of optimization, summation notation is used to find the sum of a sequence of terms. The summation below represents a summing of the first five values of a sequence of the variable x. The location of the values within the sequence are given by an index value, i in this case.

$$\sum_{i=1}^{5} x_i = x_1 + x_2 + x_3 + x_4 + x_5$$

Coefficient values may also be included in a summation, as shown below.

$$\sum_{i=1}^{5} (10 - i)\, x_i = 9x_1 + 8x_2 + 7x_3 + 6x_4 + 5x_5$$

A common mistake is to use an asterisk, *, to indicate multiplication in LaTeX.

$$\sum_{i=1}^{5} (10 - i)\, x_i = 9 * x_1 + 8 * x_2 + 7 * x_3 + 6 * x_4 + 5 * x_5$$

While the * is used for multiplication in R, it both takes more space and is not considered formal mathematical notation. In LaTeX you should instead use `\\cdot` in place of the asterisk. In other words, $\cdot$ in place of $*$. The result is the following:

$$\sum_{i=1}^{5} (10 - i)\, x_i = 9 \cdot x_1 + 8 \cdot x_2 + 7 \cdot x_3 + 6 \cdot x_4 + 5 \cdot x_5$$

When it is clearly implied by the expression as in the above example where we are multiplying a number and a variable, the `\\cdot` is optional ($9x$ vs. $9 \cdot x$). On the other hand, it is much more important when a number is not involved (Ax vs. $A \cdot x$).

B.9 Applications of Summation

Sequence applications of summation notation can be very practical in that we can extract real-world data values given in an array or a matrix.

TABLE B.2 Production Costs for Product 1

	Product 1
Design	11
Materials	12
Production	13
Packaging	14
Distribution	15

For example, let's imagine that the itemized cost for the production of a product from start to finish is that, which is given in the table below.

```
# Load library
library(kableExtra,quietly = TRUE)

# Create data table
M1 <- matrix(c(11,12,13,14,15), ncol=1)
rownames(M1)<-list("Design", "Materials",
                   "Production", "Packaging",
                   "Distribution")
colnames (M1) <- list("Product 1")
kbl (M1, booktabs=T,
        caption="Production Costs for Product 1")
```

We might wish to determine the total cost to produce the product from start to finish. We can extract the data from our cost matrix and use summation to find the total cost.

$$\sum_{i=1}^{5} x_i$$
$$= x_1 + x_2 + x_3 + x_4 + x_5$$
$$= 11 + 12 + 13 + 14 + 15$$
$$= 65$$

Let's say we now have additional products being produced.

```
# Load Data Table 2
M2<-matrix(c(11,12,13,14,15,21,22,23,24,25,31,32,33,34,35), ncol=3)
rownames(M2)<-list("Design", "Materials", "Production",
```

TABLE B.3 Itemized Production Costs for Three Products

	Product 1	Product 2	Product 3
Design	11	21	31
Materials	12	22	32
Production	13	23	33
Packaging	14	24	34
Distribution	15	25	35

```
                  "Packaging", "Distribution")
colnames(M2)<-list("Product 1", "Product 2", "Product 3")
kbl(M2, booktabs=T, caption=
        "Itemized Production Costs for Three Products")
```

We might wish to determine the cost to produce each of the three products from start to finish. We could show this with the following summation notation.

$$\sum_{i=1}^{5} x_{i,j} \quad \forall\ j$$

This notation indicates that we are summing the cost values in the i rows for each product in column j. Note that the symbol $\forall$ shown in the summation above translates to the phrase "for all." The summation expression above can be interpreted as "the sum of all values of $x_{i,j}$, starting with an initial value of $i = 1$, *for all* values of j." The expression will result in j summations.

$$\begin{aligned}
&\sum_{i=1}^{5} x_{i,j} \quad \forall\ j \\
&= x_{1,1} + x_{2,1} + x_{3,1} + x_{4,1} + x_{5,1} \\
&= 11 + 12 + 13 + 14 + 15 \\
&= 65 && \text{Cost for Product 1}
\end{aligned}$$

AND

$$
\begin{aligned}
&= x_{1,2} + x_{2,2} + x_{3,2} + x_{4,2} + x_{5,2} \\
&= 21 + 22 + 23 + 24 + 25 \\
&= 115 \qquad \text{Cost for Product 2}
\end{aligned}
$$

AND

$$
\begin{aligned}
&= x_{1,3} + x_{2,3} + x_{3,3} + x_{4,3} + x_{5,3} \\
&= 31 + 32 + 33 + 34 + 35 \\
&= 165 \qquad \text{Cost for Product 3}
\end{aligned}
$$

We can see that the summation expression resulted in three summation values since j, can take on values of 1, 2, or 3. These summation values are 65, 115, and 165, representing the total cost from start to finish to produce Product 1, Product 2, and Product 3, respectively.

B.10 Double Summation

For some projects or models, we may need to add one summation into another. This procedure is called "double summation." Consider the following double summation expression: `\sum_{i=1}^3\sum_{j=1}^4 (i+j)`

$$\sum_{i=1}^{3}\sum_{j=1}^{4}(i+j)$$

Note that the expression contains two $\sum$ symbols, indicating a double summation. The double summation would expand as shown below.

$$
\begin{aligned}
&\sum_{i=1}^{3}\sum_{j=1}^{4}(i+j) \\
&\qquad = (1+1) + (2+1) + (3+1) \\
&\qquad + (1+2) + (2+2) + (3+2) \\
&\qquad + (1+3) + (2+3) + (3+3) \\
&\qquad + (1+4) + (2+4) + (3+4)
\end{aligned}
$$

B.11 Applications of Double Summation

Consider a transportation application using double summation in which we want to ship a given amount of product X from location i to location j, denoted $X_{i,j}$. The summation notation for this application is shown below.

$$\sum_{i=1}^{n}\sum_{j=1}^{m} X_{i,j}$$

Expanding on the previous summation, we may also want to include shipping costs C from location i to location j, denoted $C_{i,j}$. Combining the amount of product with the related shipping costs would result in the double summation expression shown below.

$$\sum_{i=1}^{n}\sum_{j=1}^{m} C_{i,j}X_{i,j}$$

B.12 Exercises

Exercise B.1 (Expand-Summation1). Write all terms and calculate the summation for the following:

$$A. \quad \sum_{x=1}^{4} x+3$$

$$B. \quad \sum_{x=0}^{5} 8x-1$$

Exercise B.2 (Expand-Summation2). Write as a sum of all terms in the sequence.

$$\sum_{x=1}^{6}(2i)x$$

Exercise B.3 (Evaluate-Summation1). Write a summation to represent the total cost associated with producing three items of Product 1. Use the values from the first table in the Appendix to evaluate your summation expression.

Exercise B.4 (Evaluate-Summation2). Write all terms and calculate the summation for each exercise.

$$A. \qquad \sum_{i=1}^{3}\sum_{j=1}^{4}(i \cdot j)$$

$$B. \quad \sum_{i=1}^{5}\sum_{j=1}^{2}(3 \cdot i - j)$$

Exercise B.5 (Employee-Summation). A company compensates its employees with an annual salary and an annual bonus.

A. Write an expression using summation notation to represent the total annual compensation i (including salary and bonus) for each job title j.
B. Write a double summation expression to represent the total amount the company pays annually to compensate all its employees if each job title has n_j employees.

(Need possible sample expression for A. and B.)

C

Troubleshooting

C.1 Overview

The goal of this chapter is **not** to cover all errors that might arise while using R or RMarkdown. The goal of this Appendix is to demonstrate and discuss some common errors that arise in building optimization models using R. The error messages may not always be clear and here we provide a review. It is good practice to carefully look over the error message when you get one to see if you can find a clue as to where the problem might be.

All of these problems have stumped me or my students at some point. Whether the problems arise while doing optimization with `ompr`, creating tables to display these results, expressing the mathematical model using LaTeX, or creating the PDF for disseminating results, all of these errors can occur in other situations as well.
This appendix uses a mix of both real, working code along with images of non-working code and the errors they generate. Images are used so that this writeup can itself be properly knitted without errors.

C.2 Model Building

C.2.1 Define and Formulate before Implementing

One of the most common errors is to jump straight into coding a model into `ompr` before really understanding the model. This is the equivalent of trying to build a house without a plan. There will be a lot of wasted time and effort as well as making things difficult for anyone to assist. Algebraic linear programming models are generally only a few lines long and will often look similar to related models. Changes at this point are also very easy to make. If the model is *good* – then it is a matter of just making sure that the implementation matches the plan (formulation.)

Software engineers have a similar perspective that the cost of bug fixes goes up by a factor of 10 at every stage of development from specification development, to prototyping, to coding, to end user testing.

As an example, new modelers often get confused between which items are pieces of data to be fed into the model and which things are decision variables. Frequently, I will see people that jump too quickly to implementation going back and forth often using the same item as both data and as an optimization variable. Without this being clear, no implementation regardless of how clever, can succeed.

C.2.2 Failing to Look for Past Optimization Models

When conducting a complex optimization application, it is helpful to look at the literature. Others have likely built related models that can give a starting point for ideas on formulating. Usually some customizing is needed to fit address aspects of the new application, but it is much easier than starting with a blank sheet of paper (or screen as the case may be.) In fact, by doing this you may find implementations in algebraic modelings languages that are straightforward to translate to `ompr` as well.

C.2.3 Misrendering of PDF

Just because a PDF is knitted or rendered does not mean that it is correct. A variety of common errors often creep in such as unrendered section headings from the RMarkdown file. Consider the following text from an RMarkdown document.

```
end of sentence.
##Section Heading
Beginning of next sentence.
```

This fragment has three problems, each of which may cause the author's intended 2nd level heading to not be recognized as a heading.

1. There should be a blank line after the "end of sentence." line.
2. The pound symbols should be followed by a space before the "Section Heading."
3. There should be another blank line before the "Beginning of next sentence."

None of these will cause an error message to appear and they can easily occur by accident while editing a document, so it is just good practice to look over the PDF for this or similar issues.

C.2.4 Blank Lines in LaTeX

Put simply, blank lines are not allowed in LaTeX equations but this problem can be one of the most puzzling ones that readers have seen. The problem is that everything looks right and knits to HTML, but it doesn't knit to PDF. Part of the confusion is caused by the LaTeX processor being different from RStudio's Math rendering system, which is used for previewing in an RMarkdown document. The result is that something may be visible in the preview and look correct but not render in the PDF because of differing strictness in the implementation of parsing of the math notation.

The following screen capture shows a LaTeX formulation in an RMarkdown document that has a blank line in between the inequality constraint and the non-negativity constraints. The RStudio math preview ignores the empty line and everything looks fine.

```
$$
\begin{split}
\begin{aligned}
\text{Max:  } & 20\cdot A+14\cdot B \\
\text{s.t.: } & \\
  & 6\cdot A+2\cdot B \leq 200 \\

  & A, B \geq 0
  \end{aligned}
  \end{split}
$$
```

Max: $20 \cdot A + 14 \cdot B$

s.t.:

$6 \cdot A + 2 \cdot B \leq 200$

$A, B \geq 0$

FIGURE C.1 Preview Ignores Blank LaTeX Line.

It renders in HTML and in the RMarkdown preview without generating an error by ignoring the blank line but when knitted to PDF, a full LaTeX processor is used, which is generally more strict in enforcing LaTeX requirements. The result is that this blank line causes an error when attempting to knit to PDF and terminates.

Again, the error message may not be obvious but googling would often find a hint as to the source of the error.

If you had really intended to add a blank line before the non-negativity constraints, simply add the LaTeX linebreak code of a double slash, \\.

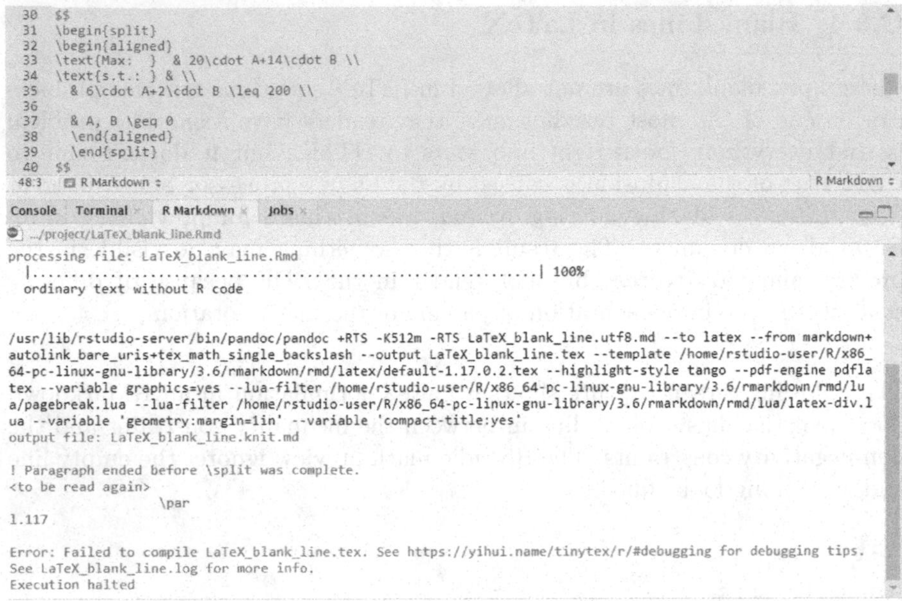

FIGURE C.2 Rendering Error Caused by Blank LaTeX Line.

$$
\begin{aligned}
\text{Max: } & 20\cdot A+14\cdot B \\
\text{s.t.: } & \\
& 6\cdot A+2\cdot B \leq 200 \\
\\
& A, B \geq 0
\end{aligned}
$$

Note that LaTeX may return a lot of warnings that do not represent problems but are provided out of an abundance of caution. For example, see the following:

```
LaTeX Warning: Command \textellipsis invalid in math mode on input line
1324.
```

The line number refer to the intermediate .tex file rather than the original .Rmd file. When LaTeX fails, it is sometimes necessary to look at both the .log and .tex files to see where the error is occurring.

C.2.5 Problems with PDF Creation

Knitting to PDF uses the LaTeX environment so a variety of other issues may arise. While sometimes these are spotted late in a project, the source is often at early stage of work so I am including it in the Model Building section.

Typically everything must be correct for an RMarkdown to be knitted. For example, when I want to rebuild this book, I will often use the `Build Book` command in RStudio. This will go through all analyses and generate a fully up to date version of the book. Alas, sometimes I only want a specific chapter, perhaps because there is an error in another chapter that I'm wrestling with. Even a single error can cause this to not build. Knitting to PDF would be next option but the various options and settings that I'm using for this book seem to prevent the Knit to PDF from working. A third option is to use the render command directly. The following command will render the file for the second chapter of the book.

```
library(rmarkdown)
rmarkdown::render("02-A_First_LP.Rmd", "pdf_document")
    # Renders as a regular PDF document
```

Another possible problem is having the PDF file open from having previously created it. This will typically generate a message about being unable to write the PDF file. The solution is to simply ensure that the PDF is not open in any other program.

A related issue is that there could be a temporary file for debugging purposes that needs to be deleted before it can be done. I often see this with a Markdown file (`.md`), not to be confused with the RMarkdown file (`.Rmd`).

Avoid using an underscore, `_`, in the code chunk label for any code chunk that involves a `kable` or graphical figure. This can cause strange errors when knitting to PDF. While the underscore is valid for a code chunk name, its importance in LaTeX causes problems – it is best to just a dash.

Some PDF errors are caused by the strict enforcement of LaTeX requirements. If the `.tex` file is generated but the `.pdf` is incomplete or missing, you can sometimes overcome these problems by using `latexmk` to rerun the Tex engine on the PDF. This is not meant to be a cure-all and is only meant for late stage errors. For example, this book sometimes requires running `latexmk OMUR.tex -pdf -f -interaction=nonstopmode` after building the `pdf_book` generates a LaTeX error at the bibliography stage. Of course tracking down the LaTeX error is preferred.

If there is a code chunk that is causing fatal errors, you can always use the code chunk option of `eval=FALSE` in order to temporarily turn off the execution while working on other parts of the PDF and model.

C.3 Implementation Troubleshooting

This section includes a variety of technical problems. Implementation troubleshooting can be tricky enough without compounding the challenge by not having a clear model. Implementation will take a lot more time and troubleshooting will be a lot harder if the definitions are not precise and the formulation is complete.

C.3.1 Errors in a Piped Model

While piping is very convenient for simplifying the implementation it has a couple of major drawbacks. First, piping may be a little slower than non-piped implementations. Second and more important for debugging problems – an error message in one line will point to the entire piped object rather than the specific line. This can make it very hard to see where the problem is.

As an example, running all code chunks for a case returned this error message to the right in my console. (Note that a similar message would also happen if I were knitting an RMarkdown document.) Also, in this example, I used the original piping operator, `%>%` which required the `magrittr` package rather than the new piping operator `|>` built into R 4.1.0 and higher. The same issue occurs regardless of the piping operator used.

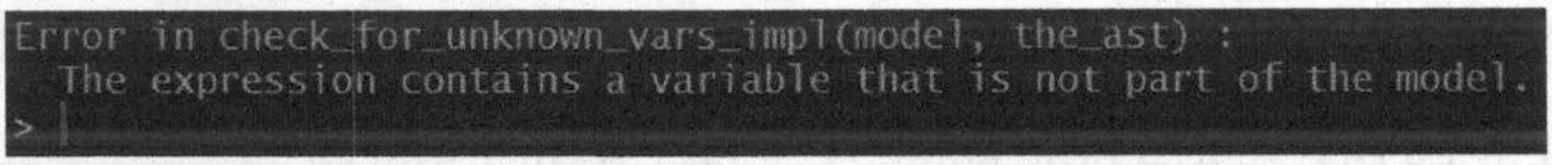

```
Error in check_for_unknown_vars_impl(model, the_ast) :
  The expression contains a variable that is not part of the model.
> |
```

FIGURE C.3 Name Conflict Error between R and `ompr`.

One of the many advantages to a modern integrated development environment, IDE, like RStudio is that it helps in debugging. In this case, looking in the code pane, RStudio has a red vertical line indicating R's best guess as to where the problem is. Unfortunately, we have a long vertical bar covering 11 lines, the entire piped object, as the source of the problem rather than a single line. When piping, it is treated as a single line and R can't narrow down the problem further. I'll leave it as a challenge for the reader to the find the error in the code to the right.

The result is that it may be helpful to focus on building up an unpiped model, while you are in active debugging.

While piping is discussed in chapter 2, it is helpful to highlight the same model side by side, implemented without piping and with piping.

```
result0 <- MIPModel() %>%
  add_variable(Ants, type = "continuous", lb = 0) %>%
  add_variable(Bats, type = "continuous",lb = 0) %>%

  set_objective(7*Ants + 12*Bats, "max") %>%

  add_constraint(1*Ants + 4*Bats<= 800) %>% #machining
  add_constraint(3*ants + 6*Bats<= 900) %>% #assembly
  add_constraint(2*Ants + 2*Bats<= 480)  %>% #testing
  add_constraint(2*Ants + 10*Bats<= 1200) %>% #sensors
  solve_model(with_ROI(solver = "glpk"))
```

FIGURE C.4 Inability to Focus on Error Source in a Piped Model.

Standard OMPR Model

```
model0 <- MIPModel()       # Initialize an empty model
model0 <- add_variable(model0, Ants,
                       type = "continuous", lb = 0)
model0 <- add_variable(model0, Bats,
                       type = "continuous",lb = 0)
model0 <- set_objective(model0, 7*Ants + 12*Bats, "max")
model0 <- add_constraint(model0, 1*Ants + 4*Bats <= 800)
                       # machining
model0 <- add_constraint(model0, 3*Ants + 6*Bats <= 900)
                       # assembly
model0 <- add_constraint(model0, 2*Ants + 2*Bats <= 480)
                       # testing
model0 <- add_constraint(model0, 2*Ants + 10*Bats <= 1200)
                       # sensors
result0 <- solve_model(model0, with_ROI(solver = "glpk"))
```

OMPR Model with Piping

```
result0 <- MIPModel() %>%
  add_variable(Ants, type = "continuous", lb = 0) %>%
  add_variable(Bats, type = "continuous",lb = 0) %>%

  set_objective(7*Ants + 12*Bats, "max") %>%

  add_constraint(1*Ants + 4*Bats<= 800) %>% #machining
  add_constraint(3*Ants + 6*Bats<= 900) %>% #assembly
  add_constraint(2*Ants + 2*Bats<= 480)  %>% #testing
  add_constraint(2*Ants + 10*Bats<= 1200) %>% #sensors
  solve_model(with_ROI(solver = "glpk"))
```

FIGURE C.5 Piped vs. Unpiped Model.

Unpiping a piped model is easy. The first step is to remove all of the pipe operators `%>%` or `|>`. Second, every `ompr` command after the model is initialized needs to be modified to name the model that is being enhanced. Lastly, every line needs to say to what the enhanced model is being assigned. The following figure is an unpiped version of the earlier code chunk that has the same error as the earlier screen capture, but it is now much easier to find the error as the reader only has to look at one line of code – the line to the right of the red vertical bar.

```
model0 <- MIPModel()       # Initialize an empty model
model0 <- add_variable(model0, Ants,
                       type = "continuous", lb = 0)
model0 <- add_variable(model0, Bats,
                       type = "continuous",lb = 0)
model0 <- set_objective(model0, 7*Ants + 12*Bats, "max")
model0 <- add_constraint(model0, 1*Ants + 4*Bats <= 800)
                       # machining
model0 <- add_constraint(model0, 3*ants + 6*Bats <= 900)
                       # assembly
model0 <- add_constraint(model0, 2*Ants + 2*Bats <= 480)
                       # testing
model0 <- add_constraint(model0, 2*Ants + 10*Bats <= 1200)
                       # sensors
result0 <- solve_model(model0, with_ROI(solver = "glpk"))
```

FIGURE C.6 Finding an Error in an Unpiped Model.

In general, since the model name or variant may occur twice in every line of an unpiped `ompr` model, keeping a short model name such as `m0` instead of `model0` would help keep the code easier to read.

C.3.2 Undefined Object in `ompr`

Let's now review the error that was generated in the previous section. This is a common `ompr` error. Of course an object must exist before it can be used in R. It might be confusing or difficult to identify the error though.

A student told me that they spent hours trying to identify the error in the following code chunk. Using a piped object might misattribute the line or the source error to the beginning of the piped command. If you see this error, read the previous section on dealing with piped objects.

The error message from `ompr` is informative.

```
Error in check_for_unknown_vars_impl(model, the_ast): The expression
contains a variable that is not part of the model.
```

The message suggests that perhaps the user forgot to create a variable using `add_variable` that was used in the objective function or a constraint. Another possibility is that the user has a problem with a data object used in the `ompr` model. The result is that users may still have difficulty finding the source of the problem.

The console reflects what I did to help in debugging. These are the steps that I followed:

- Sweep environment variables to ensure that there is nothing that might affect the analysis.
- `Run All` code chunks from the pull down menu.
- Identify the code chunk with the error.
- Check to make sure that every linear programming variable used in the model is created using the `add_variable` function. In this example, there is only one set of variables, `x[i,j]`, and it was handled correctly.
- Enter each R object in the console to see whether it is defined yet. In this case, `NSupply` was defined but when I tried `Cost` it clearly stated that it was not yet defined.

In this case, I then scrolled up and confirmed that the student had defined `NSupply`, `NDest`, and all the other data items used in the linear program except

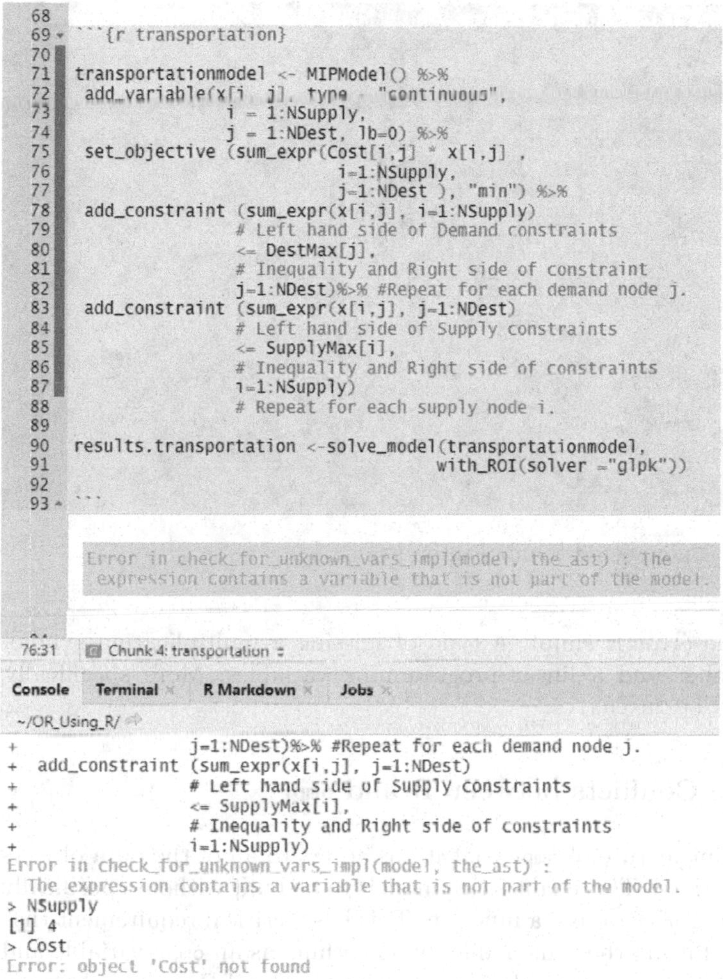

```
```{r transportation}

transportationmodel <- MIPModel() %>%
  add_variable(x[i, j], type = "continuous",
               i = 1:NSupply,
               j = 1:NDest, lb=0) %>%
  set_objective (sum_expr(Cost[i,j] * x[i,j] ,
                          i=1:NSupply,
                          j=1:NDest ), "min") %>%
  add_constraint (sum_expr(x[i,j], i=1:NSupply)
                  # Left hand side of Demand constraints
                  <= DestMax[j],
                  # Inequality and Right side of constraint
                  j=1:NDest)%>% #Repeat for each demand node j.
  add_constraint (sum_expr(x[i,j], j=1:NDest)
                  # Left hand side of Supply constraints
                  <= SupplyMax[i],
                  # Inequality and Right side of constraints
                  i=1:NSupply)
                  # Repeat for each supply node i.

results.transportation <-solve_model(transportationmodel,
                                     with_ROI(solver ="glpk"))

```

 Error in check_for_unknown_vars_impl(model, the_ast) : The
  expression contains a variable that is not part of the model.
```

```
76:31   Chunk 4: transportation
Console   Terminal   R Markdown   Jobs
~/OR_Using_R/
+                   j=1:NDest)%>% #Repeat for each demand node j.
+   add_constraint (sum_expr(x[i,j], j=1:NDest)
+                   # Left hand side of Supply constraints
+                   <= SupplyMax[i],
+                   # Inequality and Right side of constraints
+                   i=1:NSupply)
Error in check_for_unknown_vars_impl(model, the_ast) :
  The expression contains a variable that is not part of the model.
> NSupply
[1] 4
> Cost
Error: object 'Cost' not found
```

FIGURE C.7 Error from an Defined Variable.

for `Cost`. This will also help capture an inconsistent spelling such as `Cost` vs. `cost` or `Costs`.

C.3.3 Unexpected Symbol in `ompr`

Another problem for a student occurred with in the following code chunk of an `ompr` linear programming model generated an 'Error: unexpected symbol in:' message.

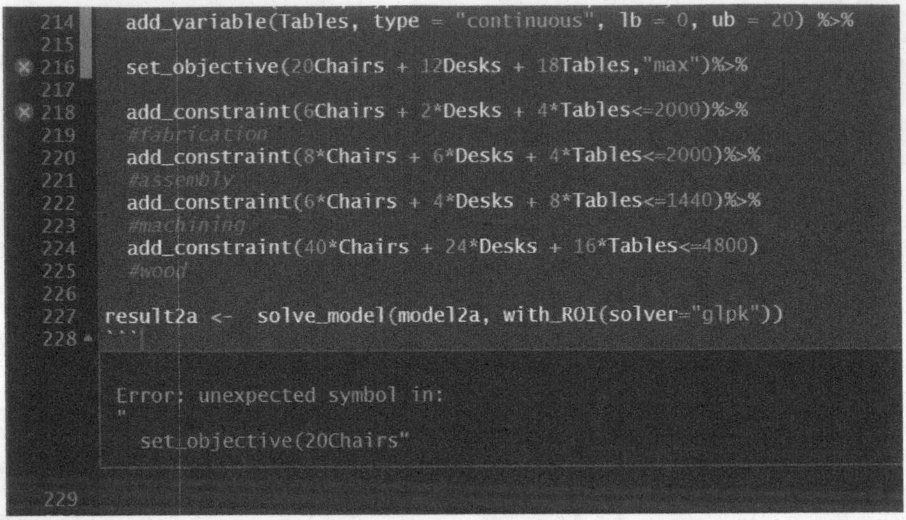

```
    add_variable(Tables, type = "continuous", lb = 0, ub = 20) %>%

    set_objective(20Chairs + 12Desks + 18Tables,"max")%>%

    add_constraint(6Chairs + 2*Desks + 4*Tables<=2000)%>%
    #fabrication
    add_constraint(8*Chairs + 6*Desks + 4*Tables<=2000)%>%
    #assembly
    add_constraint(6*Chairs + 4*Desks + 8*Tables<=1440)%>%
    #machining
    add_constraint(40*Chairs + 24*Desks + 16*Tables<=4800)
    #wood

result2a <-  solve_model(model2a, with_ROI(solver="glpk"))
```

```
Error: unexpected symbol in:
"
  set_objective(20Chairs"
```

FIGURE C.8 Unexpected Symbol in ompr.

In this case, the error is simply a typo of missing a multiplication symbol between a number and a linear programming variables. More specifically, `20Chairs` should be `20*Chairs`.

C.3.4 Name Conflicts between R and `ompr`

It is an important feature of `ompr` that it has access to all of the data objects that you have defined. This enables rich models to be built without specifically passing each piece of data into a function. This has a critical requirement that you should avoid using the same name for something as an `ompr` variable and an R object. This can happen when you want to have a LP variable be used for the same purpose in the rest of our R code or perhaps because the variable name is used elsewhere in your work for other purposes. Simple variable names such as `x` or `y` are particularly likely to find multiple uses and therefore conflicts. **Hint:** Sometimes it may be helpful to clear your environment to avoid other conflicts. This may help resolve some other inscrutable errors.

This example illustrates what happens when you have an object in your general R environment with the same name as an `ompr` model variable.

Here is a very simple linear program formulated and solved using `ompr`. Again, it uses the old piping operator instead of the new `|>` operator.

```
result1 <- MIPModel() %>%
  add_variable(A, type = "continuous", lb = 0) %>%
```

```
  add_variable(B, type = "continuous",lb = 0) %>%

  set_objective(20*A + 14*B, "max") %>%

  add_constraint(6*A + 2*B <= 200) %>%
  solve_model(with_ROI(solver = "glpk"))
```

Let's verify that it solved to optimality.

```
result1
```

```
## Status: optimal
## Objective value: 1400
```

Now, let's redo this with an identical LP, but renaming the variables from `A` and `B` to `C` and `D`. To trigger this problem, we will define `C` to be a matrix.

```
C <- matrix(c(1,2,3,4), ncol=2)
```

Now, we create the same LP but renaming `A` and `B` as `C` and `D`.}

```
result2 <- MIPModel() %>%
  add_variable(C, type = "continuous", lb = 0) %>%
  add_variable(D, type = "continuous",lb = 0) %>%
  set_objective(20*C + 14*D, "max") %>%
  add_constraint(6*C + 2*D <= 200) %>%
  solve_model(with_ROI(solver = "glpk"))
```

Since this LP is identical to the previous one other than changing names, it *should* work. Alas, we get an error message.

Notice that the error message may not actually refer to the variables causing the problem. In our example, variable `C` is causing the problem, but it refers to `x`. This is because the error is actually occurring deeper in the `ompr` code which has already transformed the actual variables into a more abstract notation.

```
```{r}
C <- matrix(c(1,2,3,4), ncol=2)
```


```{r Nonworking_LP, echo=TRUE}
result2 <- MIPModel() %>%
  add_variable(C, type = "continuous", lb = 0) %>%
  add_variable(D, type = "continuous",lb = 0) %>%
  set_objective(20*C + 14*D, "max") %>%
  add_constraint(6*C + 2*D <= 200) %>%
  solve_model(with_ROI(solver = "glpk"))
```

Error in vapply(matrices, function(constraint) { : values
 must be length 1, but FUN(X[[1]]) result is length 4
```

FIGURE C.9 Error Due to Name Conflict between R and ompr.

```
> C <- matrix(c(1,2,3,4), ncol=2)
> result2 <- MIPModel() %>%
+   add_variable(C, type = "continuous", lb = 0) %>%
+   add_variable(D, type = "continuous",lb = 0) %>%
+   set_objective(20*C + 14*D, "max") %>%
+   add_constraint(6*C + 2*D <= 200) %>%
+   solve_model(with_ROI(solver = "glpk"))
There are variables in your environment that interfere with your defined model vari
ables: C,D. This can lead to unexpected behaviour.There are variables in your envir
onment that interfere with your defined model variables: C,D. This can lead to unex
pected behaviour.There are variables in your environment that interfere with your d
efined model variables: C,D. This can lead to unexpected behaviour.Error in vapply
(matrices, function(constraint) { :
  values must be length 1,
 but FUN(X[[1]]) result is length 4
> |
```

FIGURE C.10 Error as Displayed in Console from Run All Chunks.

If you run all chunks, you may get the following error message. It doesn't make clear which variable is generating the error but does list the `ompr` variables so that you can perhaps narrow them down.

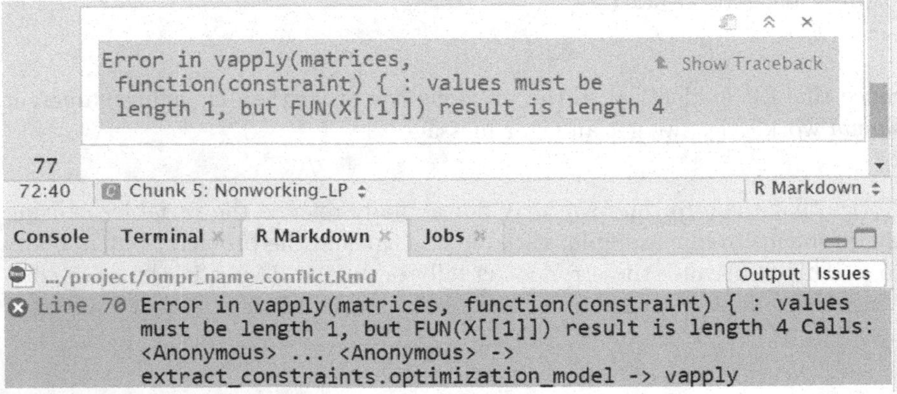

Another way to potentially deal with this problem or others is to periodically clear or sweep your RStudio environment of past objects. This can be done by using the broom icon in the Environment tab of RStudio.

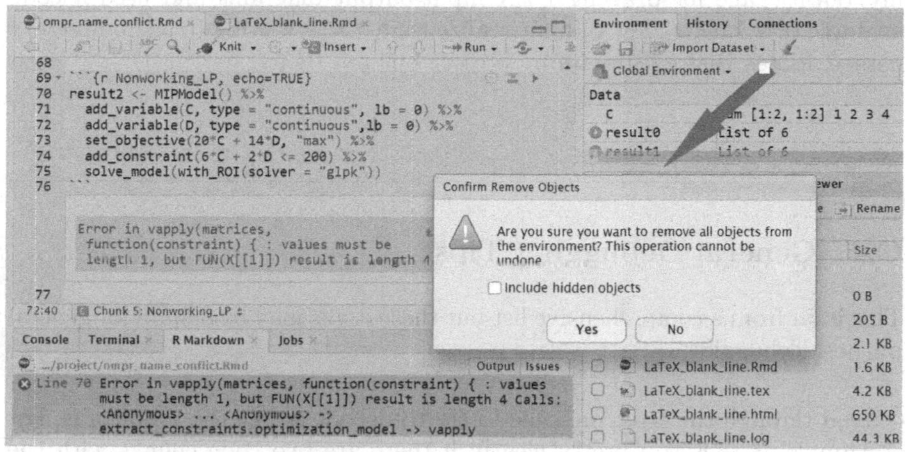

FIGURE C.11 Sweeping Out the Environment.

Since the problem is having the same name for objects inside and outside of `ompr` an an easy solution is to adopt a convention that differentiates `ompr` variables. I have adopted the convention of prefixing all `ompr` variables with a capital `V` to suggest that it is a mathematical programming variable. Readers can then easily differentiate which items in constraints or the objective function are variables (specific to `ompr`) and which are sources of data from outside of `ompr`. The resulting model is shown below. When expressing the model mathematically for readers, I can then omit the `V` prefix.

```
# Fixing the Problem by Giving Variables Unique Names

result2 <- MIPModel() %>%
  add_variable(VC, type = "continuous", lb = 0) %>%
  add_variable(VD, type = "continuous", lb = 0) %>%
  set_objective(20*VC + 14*VD, "max") %>%
  add_constraint(6*VC + 2*VD <= 200) %>%
  solve_model(with_ROI(solver = "glpk"))
```

C.3.5 Blindly Reusing Code

You will often find examples of code that is helpful. It is important to try to read what the code is doing before using it. For example, a reader tried using my `TRA` package for drawing DEA input-output diagrams and used a code example that I had. They did not realize that it was writing to a subdirectory named `images` that existed in my project but not theirs.

C.4 General Debugging Tips

This is far from a comprehensive list but these items may be helpful for general purpose debugging.

- Read through the error message to try to decode where the problem is. For example, it tells you pretty clearly if there are two code chunks with the same name.
- Check for spelling, capitalization, or singular/plural errors in variable names. The human reader's eye tends to be more forgiving of inconsistencies than computers.
- Try stepping through the code by evaluating one code chunk (or even line) at a time to narrow down where an error may be occurring.
- Use the console to examine run code and display variable values directly.
- Do a web search for the error message, leaving out items specific to your code such as variables or code chunk names.
- Narrow down where to look for problems by line numbers or code chunk names.
- Check your parentheses carefully – it is easy to get a parenthesis misplaced.
- If you are getting a LaTeX error, look at the TeX log file. You can scroll down to the very end to see where things went wrong. This message may give more clues to where the error arose.
- See if you can create a "minimal" example that reproduces the same problem in a new file. This can be very helpful for others to see where your problem is and not read through many lines of working code. As you trim it down in a second file, you might also have a Eureka moment where you see it yourself.
- If you have a complex statement that does three or four things in one line, try working from inside out by running parts of the command in the console to make sure that they each generate the results that you intend.
- Inspect the data that you are using. Perhaps the data is not of the right form.
- Caching analysis results in RStudio can be helpful for advanced users with

computationally demanding models but can result in subtle problems. Avoid it unless you plan to be very careful.

- If you are using piping, either the new `|>` pipe operator built into R, or the older `%>%` pipe, make sure to not use a pipe at the very end of the series of commands being connected.

C.5 Getting Help

After going through the above ideas, you may still be stuck. We've all been there before in working with computers. Given the large number of people using R, it is likely that many other people have had the same problem before so a good search will probably find a discussion of the problem and resolution. I often use google to search for the generalized text snippets of error messages, leaving out things unique to my case such as a variable name. Sometimes add in the words "R", "package" or the specific package name used as appropriate. Using the search qualifiers of `+` or `-` can be very helpful in finding specific items. For example a search of `boxplot +r -python` will find pages that discuss boxplots and R but exclude any page that mentions python.

After going through these efforts, you may still not find a solution. The R community is a great resource for getting answers to problems but there is a common practice that is important to emphasize. When you ask for help, you are typically requesting a person that you don't know to review your work for free.

Try to come up with a simple, short reproducible example. The goal is to make it as quick and easy for the person to examine this. Make it as short and simple as possible, remember that depending upon where you are asking for help, they may not be familiar with linear programming or the particular packages that you are using. For example, if you have a problem building an ompr model, try removing as many constraints as possible while still getting the error. The person trying to help won't care about the other missing constraints. You can always add them back later.

Another tip for making the reproducible example easier to use simplified variable names. Again, the reader doesn't really care about your specific model but this makes it shorter and easier to read. Of course, this might cause (or fix?) name conflicts or collisions.

A detailed discussion of how to create good reproducible examples can be found online.
`https://reprex.tidyverse.org/articles/reprex-dos-and-donts.html`

Including the code in a way that can be easily run is often helpful. If you are using rstudio.cloud, you may even create a simple project that someone can jump into for debugging purposes. I've done this with students, but it would not work well for posting a link into an open Internet forum.

If you have posted the question or emailed someone and there were suggestions given, a response is usually appreciated or in a forum post, may be helpful for a future person that has the same problem.

D

Making Good Tables

D.1 Importance of Tables in Modeling

```
library(knitr)  # Only two packages used in this Appendix
library (kableExtra)
knitr::opts_chunk$set(echo = TRUE)
```

The data scientist may interpret "A picture is worth a thousand words" as "A figure or table may be worth a thousand words." The most elegant and sophisticated analysis is useless if the results are not documented or used. While figures are a subject too broad for an appendix, tables are so core to analysis that it warrants explaining how we created the tables for the book. This Appendix is provided to show all the techniques that were used in making the tables throughout the book to enable the reader to use them as well. Providing the information in one place in the Appendix will reduce the repetition in the other chapters.

Throughout the bulk of the book, sometimes the command for generating the table is shown for demonstrating particular techniques, but they are discussed in more detail in this appendix. The RMarkdown source documents have the code for each table in the book.

Because of the importance of tables, many packages have been developed for creating rich and sophisticated tables such as knitr's `kable`, `pander`, `grid.table`, `huxtable`, `flextable`, `gt`, `xtable`, `ztable`, and more. Previous versions of this book extensively used `pander` and `grid.table`. After a lot of trial and error, I settled on knitr's `kable` using the enhancements provided by `kableExtra` for the following reasons:

- Ease of use
- Widely used
- Well documented, including use with `bookdown`
- Straightforward table resizing

DOI: 10.1201/9781003051251-D

- Ability to use LaTeX in row and column names
- Relatively compact syntax

Just as `ggplot` implemented a widely used and powerful grammar of graphics, some of the recent packages have taken a similar approach and philosophy, specifically `huxtable`, `flextable`, and gives rise to the name of the recent `gt` from team at RStudio. These are powerful packages worth considering in the future but for the time-being, I will focus on `kable` as assisted with `kableExtra`.

There are more options in `kable` and `kableExtra` than we are covering here but this covers everything that was used in the book.

D.2 Kable vs. Kbl

In theory, everything done with `kableExtra` could be done through just `kable` since `kableExtra` serves as a friendly interface to `kable`. Actually doing everything directly through `kable` would be much more difficult and prone to user error so `kableExtra` is recommended. The first item to note is `kableExtra` provides a shorthand function, `kbl`, similar to knitr's native `kable`. In fact, it is essentially a wrapper that provides several useful benefits:

- automatic format recognition rather than requiring setting format="latex",
- provides direction options for important `kable` options which makes the command more readable and provides autocomplete, and
- shortens every line by 2 letters helping fit commands within a single printed line.

We will use `kbl` throughout this Appendix and book. If you are getting an error that `kbl` function is not recognized, you will need to load the `kableExtra` package.

D.3 Table Footnotes with Kable

Footnotes can be text or listed with demarcations for numbers, letters, or symbols.

```
m <- matrix(c(1:4),nrow=2, ncol=2)
rownames(m) <- c("Row name 1", "Row name 2")

kbl (m, booktabs=T,
     caption="Footnotes in Tables Using kbl",
     col.names=c("C1", "C2"),
     row.names=F) |>
  kable_styling(latex_options = "hold_position") |>
  footnote(general = "General comments about the table. ",
       number = c("Footnote 1; ", "Footnote 2; "),
       alphabet = c("Footnote A; ", "Footnote B; "),
       symbol = c("Footnote Symbol 1; ", "Footnote Symbol 2"))
```

TABLE D.1 Footnotes in Tables Using kbl

C1	C2
1	3
2	4

Note:
General comments about the table.
[1] Footnote 1;
[2] Footnote 2;
[a] Footnote A;
[b] Footnote B;
[*] Footnote Symbol 1;
[†] Footnote Symbol 2

D.4 Setting Row and Column Names in Kable

Kable has a handy option for changing `col.names` in the function directly. This is helpful because often the data structure has a specific naming convention for specific reasons but are not in very human readable format. Rather than always changing the column names before displaying the table or creating a new table with just different column names, this gives you the option just changing it for display purposes without affecting the original table.

One tricky thing is that while column names can be changed in `kable`, row names cannot be changed in `kable`, despite the options looking equivalent `col.names=` vs. `row.names=`.

The following code chunk seems like it should work but `row.names` generates an error because it wants a `TRUE` or `FALSE` (or their shorthand equivalents of `T` and `F`) rather than the row names. The code chunk was set to `eval=FALSE` in order to avoid causing an error. The lesson is that row names are not always handled in R in the same way as column names or as even consistently in different settings.

```
# eval=FALSE to avoid breaking error.
m <- matrix(1:4,nrow=2, ncol=2)

kable (m,
     col.names=c("C1", "C2"),
     row.names=c("R1", "R2"))
# Works for col.names but not row.names
# The following demonstrates use of row.names
```

Since `col.names` works well, there is no reason to spend more time on it. Let's instead show how to replace row names for a table. The following example gives bad row names to a matrix that I want to replace and use with an better set of row names. It then generates four tables showing these situations. The last two options give good results.

```
m <- matrix(1:4, 2, 2)

BR <- c("Bad row name 1", "Bad row name 2")
GR <- c("Good row name 1", "Good row name 2")

row.names (m)<-BR  # Treat as default to be avoided

kbl (m, booktabs=T, caption="Retain Row Names by Using `row.name=T`",
     col.names=c("C1", "C2"), row.names=T)                            |>
  kable_styling (latex_options = "hold_position")
```

TABLE D.2 Retain Row Names by Using 'row.name=T'

	C1	C2
Bad row name 1	1	3
Bad row name 2	2	4

```
kbl (m, booktabs=T, caption="Drop Row Name by Using `row.name=F`",
    col.names=c("C1", "C2"), row.names=F)                              |>
  kable_styling (latex_options = "hold_position")
```

TABLE D.3 Drop Row Name by Using ‘row.name=F‘

C1	C2
1	3
2	4

```
kbl (cbind(as.matrix(GR),m), booktabs=T,
      caption="Adding Row Names as a Leading Column",
    col.names=c("", "C1", "C2"), row.names=F)                          |>
  kable_styling (latex_options = "hold_position")
```

TABLE D.4 Adding Row Names as a Leading Column

	C1	C2
Good row name 1	1	3
Good row name 2	2	4

```
rownames(m)<-GR
kbl (m, booktabs=T,
      caption="Replacing row names in matrix before kable",
    col.names=c("C1", "C2"), row.names=T)                              |>
  kable_styling (latex_options = "hold_position")
```

TABLE D.5 Replacing Row Names in Matrix Before kable

	C1	C2
Good row name 1	1	3
Good row name 2	2	4

The `kable_styling` command has a parameter, `latex_options` that can be passed a lot of different parameters that only apply to the output of `latex`. I often use the `"hold_position"` option to encourage LaTeX to keep the table placement near the code chunk. If `"hold_position"` is not passed, LaTeX will float the table to what it considers a good place relative to the rest of the text that might be well away from the actual intent.

To summarize:

- `col.names` from `kable` makes it very easy to temporarily change the column names just for the purpose of the table.
- Row names can be added using a `cbind` to pre-pend the row names as the first column. Note that this means that an additional column name needs to be added. Here I add `""` to make it a blank column name.
- Kable treats `row.names` as a simple flag that can be set to show the original rownames `TRUE` (or `T`) and `FALSE` (or `F`) to hide the original rownames.

D.5 Booktabs vs. Default

The `booktabs` option is used to provide the formatting commonly used in books and journals. Notably it reduces the extra border lines for every element that leaves an otherwise nice table looking like it is an Excel screen capture.

```
kbl (cbind(as.matrix(GR),m),
     caption="Default format using kable.",
     col.names=c("", "C1", "C2"),
     row.names=F) |>
  kable_styling(latex_options = "hold_position")
```

TABLE D.6 Default Format Using kable

	C1	C2
Good row name 1	1	3
Good row name 2	2	4

```
kbl (cbind(as.matrix(GR),m), booktabs=T,
     caption="Kable using booktabs.",
     col.names=c("", "C1", "C2"),
     row.names=F)|>
  kable_styling(latex_options = "hold_position")
```

TABLE D.7 Kable Using booktabs

	C1	C2
Good row name 1	1	3
Good row name 2	2	4

D.6 Using LaTeX in Kable Column Names

Some applications require richer notation than simple plain text. With a little extra effort, you can get full LaTeX notation in tables. The primary use is likely to be in row and column names. As discussed earlier, column names can be set from within the `kbl` function call. Two important things to account for:

- Set `escape=F` to allow using LaTeX,
- Any command requiring a slash character in LaTeX requires a double slash. For example, `$\theta^{CRS}$` to create θ^{CRS} would instead require `$\\theta^{CRS}$` as shown below.

```
m <- matrix(1:4, 2, 2)
GR <- c("$\\phi$", "$\\omega$")

kbl(cbind(as.matrix(GR),m), booktabs=T, escape=F, caption=
    "Using LaTeX in Row and Column Names",
    col.names=c("","$\\theta^{CRS}$", "$\\lambda_A$"),
    row.names=F) |>
  kable_styling(latex_options = "hold_position")
```

TABLE D.8 Using LaTeX in Row and Column Names

	θ^{CRS}	λ_A
ϕ	1	3
ω	2	4

D.7 Fitting Tables to Page Width

Large tables can be problematic. Kable can scale a table to fit the page width easily using the `scale_down` option in `kableExtra` via the `kable_styling` and `latex_options`.

```
NWeeks <- 24
mbig <- matrix(1:(2*NWeeks), ncol= NWeeks, byrow=TRUE,)
rownames (mbig) <- c("Widgets", "Gadgets")

kbl(mbig,booktabs=T,
    caption="Scaling Table to Fit Page")|>
  kable_styling(latex_options =
                  c("hold_position", "scale_down"))
```

TABLE D.9 Scaling Table to Fit Page

Widgets	1	2	3	4	5	6	7	8	9	10	11	12	13	14	15	16	17	18	19	20	21	22	23	24
Gadgets	25	26	27	28	29	30	31	32	33	34	35	36	37	38	39	40	41	42	43	44	45	46	47	48

A few things to note:

- Multiple options can be passed at once to `latex_options` as shown above.
- The `scale_down` option is done through `latex_options` highlighting that it only works for PDFs generated through LaTeX.
- When shrunk, the font may be too small to read.
- The `scale_down` option will also scale up a table that does not fit the full page width which may make a small table look cartoonishly large.
- Some oversized tables may benefit from using the `kbl` option of `longtable=T`.

On the other hand, sometimes this is a sign that the display of this information needs to be rethought: - transposing the table to be taller rather than wider, - not displaying certain columns or rows with less information value, - using a figure of some form rather than a table, - turning the table (or page) sideways, - sticking the table into an appendix or as a file for download, or - providing a summary table of means, standard deviations, or other information.

Bibliography

Adler, J. (2012). *R in a Nutshell.* O'Reilly, Sebastopol, CA.

Anderson, T. R., Daim, T. U., and Lavoie, F. F. (2007). Measuring the efficiency of university technology transfer. *Technovation*, 27(5):306–318. WOS:000246436900006.

Baker, K. (2015). *Optimization Modeling with Spreadsheets.* Wiley, Hoboken, 3rd edition.

Banker, R., Charnes, A., and Cooper, W. (1984). Some models for estimating technical and scale inefficiencies in data envelopment analysis. *Management Science*, pages 1078–1092.

Bogetoft, P. and Otto, L. (2013). *Benchmarking with DEA, SFA, and R.* Springer, New York u.a., 2011 edition.

Charnes, A., Cooper, W. W., and Rhodes, E. (1978). *A Data Envelopment Analysis Approach to Evaluation of the Program Follow Through Experiment in U.S. Public School Education.* Division of Research Graduate School of Business Administration Harvard University, Boston.

Hillier, F. and Lieberman, G. (2020). *Introduction to Operations Research.* McGraw-Hill Education, Dubuque, 11th edition.

Kabacoff, R. (2011). *R in Action: Data Analysis and Graphics with R.* Manning ; Pearson Education [distributor], Shelter Island, NY; London.

Ragsdale, C. (2017). *Spreadsheet Modeling & Decision Analysis: A Practical Introduction to Business Analytics.* Cengage Learning, Boston, MA, 8th edition.

Rhodes, E. L. (1978). *Data Envelopment Analysis and Approaches For Measuring the Efficiency of Decision-Making Units With an Application to Program Follow-Through in U.S. Education.* PhD thesis.

Tovey, C. A. (2021). *Linear Optimization and Duality: A Modern Exposition.*

Winston, W. (2003). *Operations Research: Applications and Algorithms.* Cengage Learning, Belmont, Calif, 4th edition.

Index

Algebraic model, 39, 40
Allocation model, 50

Benchmarking package, 88
Blending constraint, 47

cbind function, 232
Column duals, *see* Reduced costs of variables
Constraint satisfaction, 185
Constraints, 8
Conventions, 6
Covering model, 51
CRAN, 4

Data envelopment analysis, 85–124
    Input-oriented, 96
    Multiplier weights, 113
    Orientation, 96, 121
    Output-oriented, 96
    Returns to scale, 100–121
    Slacks, 116
DEA, *see* Data envelopment analysis
dea.plot, 89
Decision making unit, 90
Decision variables, 8
Descriptive models, 3
Deviational variables, 210
dimnames, 227
DMU, *see* Decision making unit
DrawIOdiagram, 86

Explicit model, 39

Forall, 40

glpk, 63
Goal programming, 205–223
    Preemptive Goal Programming, 205

Heuristic methods, 34

Infeasible, 25, 26
Integer programming
    Big M, 149
    Binary variables, 143
    Branch and bound, 131–141
    Fixed charge, 147–156
    Linking constraint, 149
Interior points algorithm, 34
Inventory management, 59

Julia, 5

kable, 228, 264–268
    booktabs, 228, 268
    caption, 228
    Column and row names, 265
    digits, 110
    footnotes, 264
    hold_position, 228, 268
    kable_styling, 228
    kableExtra, 228
    kbl, 228, 264
    LaTeX in column names, 96, 269
    scale_down, 270

LaTeX, 235–245
    Blank line, 249
    latexmk, 251
    TeX log file, 260
Linear programming, 9
    Explicit formulation, 14, 18, 33
    Standard form, 60
LP, *see* Linear programming

lpSolve, 65
lpSolve package, 63
lpsolveAPI, 63

Management science, 1
Minimax, 211, 215
MIP, *see* Mixed integer programming
MIPModel|seeompr
    MIPModel, 15
Mixed integer programming, 15
Multiple optima, 26

NLP, *see* Nonlinear programming
No feasible solution, *see* Infeasible
Non-binding constraint, 31
Nonlinear programming, 9

Objective function, 8
ompr
    add_constraint, 16
    add_variable, 15
    MIPModel, 15
    Misreporting of status, 32, 176
    objective_value, 22
    Prefixing variables, 43
    R and ompr name conflicts, 43
    set_objective, 16
Operations research, xx, 1
Optimization, xx, 3

Pareto optimal, 212
Piping, 6, 20, 21, 229
    Debugging a piped model, 252
    New vs. old operator, 252
    Unpiping, 253
poscol function, 110
Production planning, 59
Python, 5

rbind function, 232
Reduced costs of variables, 77
Redundant constraint, 30
Rglpk, 65
RHS, *see* Right hand side
Right hand side, 19
RMarkdown
    Caching, 180, 261
    Inline, 45, 237
    LaTeX, 237
ROI package, 14, 63
Row duals, *see* Shadow prices
RSymphony, 65

Shadow prices, 73, 113
Simplex method, 34, 77, 130, 174, 208
Simulation, 3
Slack variable, 61
symphony, 63

TRA package, 85, 110
Transportation model, 53
Transshipment model, 57

Unbounded solution, 31

Printed in the United States
by Baker & Taylor Publisher Services

Printed in the United States
by Baker & Taylor Publisher Services